MATHEMATICAL QUICKIES

MATHEMATICAL QUICKIES

270 Stimulating Problems with Solutions

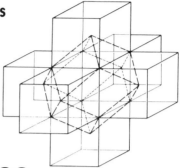

by CHARLES W. TRIGG

Dean Emeritus and Professor Emeritus
Los Angeles City College

DOVER PUBLICATIONS, INC., NEW YORK

This Dover edition, first published in 1985, is an unabridged and corrected republication of the work first published by McGraw-Hill Book Company, New York, in 1967 under the title *Mathematical Quickies*.

Manufactured in the United States of America
Dover Publications, Inc., 31 East 2nd Street, Mineola, N.Y. 11501

Library of Congress Cataloging in Publication Data

Trigg, Charles W.
  Mathematical quickies.

  Includes index.
  1. Mathematics—Problems, exercises, etc.   I. Title.
QA43.T66   1985   510'.76   85-7058
ISBN 0-486-24949-2 (pbk.)

PREFACE

In this assemblage of problems, the emphasis is upon the method of *solution*. Thus they provide a double challenge to the reader—to solve the problems and to devise solutions more elegant than the ones provided.

My collection of Mathematical Quickies began in March, 1950, when, as editor of the Problems and Questions department of *Mathematics Magazine*, I introduced a subdepartment entitled "Quickies." Its heading stated, "From time to time this department will publish problems which may be solved by laborious methods, but which with proper insight may be disposed of with dispatch. Readers are urged to submit their favorite problems of this type, together with the elegant solution and the source, if known." The Quickies section caught on at once and has retained its popularity.

An elegant *solution* is generally considered to be one characterized by clarity, conciseness, logic, and surprise. Brevity is not to be achieved by omitting steps essential for easy understanding nor by resorting to that mathematical evasion—"obviously." Naturally, unexpectedness vanishes if the reader is already familiar with the result.

Quickness and elegance are relative matters, they are not absolute. Often a seldom-used theorem or one from an advanced or different discipline will provide the magic touch. Or, it may be that mathematics of a more elementary nature will speed up the solution. A special device or one apparently unrelated to the problem may provide the quickness sought.

Good problems are likely to become anonymous over the years, sometimes due to variations in wording that disguise unacknowledged borrowing. Not to perpetuate this unfortunate situation, but recognizing that to locate with certainty the first appearance of a problem is practically impossible, in this volume reference to the author and source of the *solution* is given. Thus the appendage, "Leo Moser, *M.M.*, 25 (May, 1952), 290" to the solution of problem 15, means that the *solution* by Leo Moser has been taken from page 290 of the May 1952 issue of *Mathematics Magazine*, volume 25. If no author is given, the solution is my own. If no printed source is indicated, the author of the *solution* conveyed it directly to me.

Abbreviations employed are: *A.M.M.* (*American Mathematical Monthly*), *M.M.* (*Mathematics Magazine*), *N.M.M.* (*National Mathematics Magazine*), *P.M.E.J.* (*Pi Mu Epsilon Journal*), and *S.S.M.* (*School Science and Mathematics*). The publishers of these magazines and of *The Pentagon* were gracious enough to grant permission to reproduce material from their journals. To them and to the other sources which are clearly attached to the pertinent solutions, grateful acknowledgment is made.

Upon occasion, the original solution has been paraphrased without modifying the method. In some cases where well-known symbols facilitating compactness are not available, the exposition of the solution may seem long although the basic ideas and steps involved are relatively few.

Since an essential part of problem solution is to settle upon the particular mathematical discipline to be used, classification of the problems by mathematical field has been avoided. In this purposeful disarray, the difficulty of the problems has been varied, with relatively easy ones randomly distributed throughout the book.

To facilitate ready passage from problem to solution and vice versa, a dictionary style of heading has been provided on the pages. The problems and solutions are given in the same order so that they may be located readily.

All the ideas involved in the solutions should be comprehensible to the good high school student. Some of the problems may well provide a definite challenge to the graduate student. Should any reader find a solution more elegant than the one given, or publication of the "quickie solution" previous to the recorded reference, he is hereby asked to share the discovery with me.

2404 Loring Street   *Charles W. Trigg*
San Diego
California 92109

CONTENTS

CHALLENGE PROBLEMS

270 intriguing problems of varying difficulty have been selected from various mathematical fields and sources. They offer a double challenge—to solve the problem and to devise a neater, quicker, more elegant solution than the one published in the solution section which begins on page 77.

Problems from the different mathematical fields are distributed at random throughout the collection. No attempt has been made to isolate the easier ones. They may appear any place in the list.

The challenge problems are numbered consecutively for easy reference. Numbers of the problems printed on a page appear at the upper outside corner of the page.

1. The Careless Mailing Clerk

After a typist had written ten letters and had addressed the ten corresponding envelopes, a careless mailing clerk inserted the letters in the envelopes at random, one letter per envelope. What is the probability that exactly nine letters were inserted in the proper envelopes?

2. The Pythagorean Theorem

Prove that the square of the hypotenuse of a right triangle equals the sum of the squares of the other two sides.

3. Four Equations in Four Unknowns

Completely solve the following system of equations:

$$x + y + z + w = 10$$

$$x^2 + y^2 + z^2 + w^2 = 30$$

$$x^3 + y^3 + z^3 + w^3 = 100$$

$$xyzw = 24$$

4. A Test with Zero Score

On a 26-question test, five points were deducted for each wrong answer and eight points were credited for each correct answer. If all the questions were answered, how many were correct if the score was zero?

5. Ptolemy's Theorem

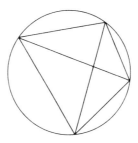

In any cyclic convex quadrilateral, show that the product of the diagonals equals the sum of the products of the opposite sides.

6. Easy Factoring

Without grouping, factor

$$x^8 - x^7y + x^6y^2 - x^5y^3 + x^4y^4 - x^3y^5 + x^2y^6 - xy^7 + y^8.$$

7. Mental Computation

Square 85 mentally.

8. A Quartic Equation

How many negative roots does the equation

$$x^4 - 5x^3 - 4x^2 - 7x + 4 = 0 \quad \text{have?}$$

9. A Million on Each Side

An array of two million points is completely enclosed by a circle having a diameter of 1 inch. Does there exist a straight line having exactly one million of these points on each side of the line? If so, why?

10. An Infinitude of Primes

Show that there is an infinitude of prime numbers.

11. Complex Numbers

Simplify $(27 + 8i)/(3 + 2i^3)$.

12. Overlapping Areas

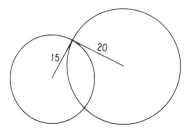

A circle of radius 15 intersects another circle, radius 20, at right angles. What is the difference of the areas of the nonoverlapping portions?

13. A Tennis Tournament

There are n players in an elimination-type singles tennis tournament. How many matches must be played (or defaulted) to determine the winner?

14. Summing a Factorial Series

Find the sum of the series

$$1(1!) + 2(2!) + 3(3!) + \cdots + n(n!).$$

15. Intersecting Cylinders

The axes of symmetry of two 2-inch right circular cylinders intersect at right angles. What volume do the cylinders have in common?

16. Representations of an Integer

The number 3 can be expressed as a sum of one or more positive integers in four ways, namely as 3, $1 + 2$, $2 + 1$, and $1 + 1 + 1$. Show that any positive integer n can be so expressed in 2^{n-1} ways.

17. Quartic with a Rational Root

Show that the quartic equation

$$(\ \ )x^4 + (\ \ )x^3 + (\ \ )x^2 + (\ \ )x + (\ \ ) = 0$$

where the gaps are filled in by any arrangement of the numbers 1, −2, 3, 4, −6, always has a rational root.

18. Resistance of a Cube

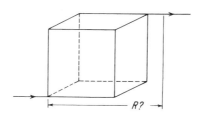

Each edge of a cube is a 1-ohm resistor. What is the resistance between two diagonally opposite vertices of the cube?

19. A Facetious Division

Each letter in the cryptarithm $A\,H\,H\,A\,A\,H/J\,O\,K\,E = H\,A$ uniquely represents a digit in the decimal scale. What is the arithmetic division?

20. The Flower Salesman

A girl entered a store and bought x flowers for y dollars (x and y are integers). When she was about to leave, the clerk said, "If you buy ten more flowers I will give you all for \$2, and you will save 80 cents a dozen." Find x and y.

21. Relative Polygonal Areas

An equilateral triangle and a regular hexagon have equal perimeters. What is the ratio of their areas?

22. The Inverted Cups

It is desired to invert the entire set of n upright cups by a series of moves in each of which $n - 1$ cups are turned over. Show that this can always be done if n is even, but never if n is odd.

23. The End of the World

On April 1, 1946, the *Erewhon Daily Howler* carried the following item: "The famous astrologer and numerologist of Guayazuela, the Professor Euclide Paracelso Bombasto Umbugio, predicts the end of the world for the year 2141. His prediction is based on profound mathematical and historical investigations. Professor Umbugio computed the value of the formula

$$1492^n - 1770^n - 1863^n + 2141^n$$

for $n = 0, 1, 2, 3$, and so on, up to 1945, and found that all the numbers which he obtained in many months of laborious computation are divisible by 1946. Now, the numbers 1492, 1770, and 1863 represent memorable dates: The Discovery of the New World, the Boston Massacre, and the Gettysburg Address. What important date may 2141 be? That of the end of the world, obviously."

Deflate the professor! Obtain his result with little computation.

24. Six Distinct Integers

Find the smallest set of six distinct integers with the property that the product of any five of them is one or more periods of the unit repetend of the remaining one. For example: $\frac{1}{41} = 0.0243902439\cdots$, so 02439 is the unit repetend of 41.

25. Length of a Helix

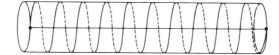

Ten turns of a wire are helically wrapped around a cylindrical tube with outside circumference 4 inches and length 9 inches. The ends of the wire coincide with the ends of the same cylindrical element. Find the length of the wire.

26. A Minimum Problem

Show that for all positive values of p, q, r, and s,

$$\frac{(p^2 + p + 1)(q^2 + q + 1)(r^2 + r + 1)(s^2 + s + 1)}{pqrs}$$

cannot be less than 81.

9

27. Digits of a Square Number

Show that any perfect square in the decimal system which has two or more digits contains at least two distinct digits.

28. Interlocking Committees

There are 15 men on the board of directors of a large company, and 20 committees are associated with the board. It is required to set up these committees so that: (1) Each board member shall belong to four committees; (2) Each committee shall have three board members; (3) No two committees shall have more than one board member in common.

29. The Jigsaw Puzzle

In assembling a jigsaw puzzle, let us call the fitting together of two pieces a "move," independently of whether the pieces consist of single pieces or of blocks of pieces already assembled. What procedure will minimize the number of moves required to solve an n-piece puzzle? What is this minimum number?

30. A Remainder Problem

Given that $f(x) = x^4 + x^3 + x^2 + x + 1$, find the remainder when $f(x^5)$ is divided by $f(x)$.

31. A Skeleton Division

Our good friend and eminent numerologist, Professor Euclide Paracelso Bombasto Umbugio, has been busily engaged testing on his desk calculator the $81 \cdot 10^9$ possible solutions to the problem of reconstructing the following exact long division in which digits indiscriminately were each replaced by x.

$$
\begin{array}{r}
x\,x\,8\,x\,x \\
x\,x\,x\,)\overline{x\,x\,x\,x\,x\,x\,x\,x} \\
\underline{x\,x\,x} \\
x\,x\,x\,x \\
\underline{x\,x\,x} \\
x\,x\,x\,x \\
\underline{x\,x\,x\,x}
\end{array}
$$

Deflate the Professor!   That is, reduce the possibilities to $(81 \cdot 10^9)^0$.

32. Triangles in a Circle

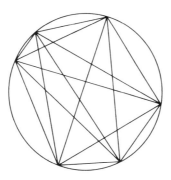

If n points on the circumference of a circle are joined by straight lines in all possible ways, and no three of these lines meet at a single point inside the circle, find the number of triangles formed, all of whose vertices lie inside the circle.

33. Easy Multiplication

Multiply 5,746,320,819 by 125.

34. A Repetitive Series

What is the nth term of the repetitive series

$$-4 + 7 - 4 + 7 - 4 + 7 - \cdots?$$

35. Radical Simplification

Simplify $\sqrt[3]{2 + \sqrt{5}} + \sqrt[3]{2 - \sqrt{5}}$.

36. The Buried Treasure

A pirate decided to bury a treasure on an island near the shore of which were two similar boulders A and B and, farther inland, three coconut trees C_1, C_2, C_3. Stationing himself at C_1, the pirate laid off C_1A_1 perpendicular and equal to C_1A and directed outwardly from the perimeter of triangle AC_1B. He similarly laid off C_1B_1 perpendicular and equal to C_1B and also directed outwardly from the perimeter of triangle AC_1B. He then located P_1, the intersection of AB_1 and BA_1. Stationing himself at C_2 and C_3, he similarly located points P_2 and P_3, and finally buried his treasure at the circumcenter of triangle $P_1P_2P_3$.

Returning to the island some years later, the pirate found that a big storm had obliterated all the coconut trees on the island. How might he find his buried treasure?

37. Bonus Payments

A company offered its 350 employees a bonus of $10 to each male and $8.15 to each female. All the females accepted, but a certain percentage of the males refused to accept. The total bonus paid was not dependent upon the number of men employed. What was the total amount paid to the women?

38. Product Simplification

Simplify the product

$$(3^{2^0} + 1)\,(3^{2^1} + 1)\,(3^{2^2} + 1)\cdots(3^{2^n} + 1)$$

39. Untangled Strings

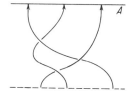

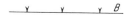

Three strings are tied to three pegs on board A. They are tangled and interwoven as shown in the figure. It is required to tie three other strings to the three free ends and attach the free ends of the new strings, which may be tangled, to the three pegs of board B in such a way that the resultant entanglement can be combed out to give three loosely parallel strings. How can this be done?

13

40. A Difference Equal to a Quotient

Find two numbers whose difference and whose quotient are each equal to 5.

41. The Dozing Student

A student awoke at the end of a class in algebra one morning just in time to hear his teacher say, " . . . and I will give you the hint that all the roots are real and positive." Looking at the board he discovered a 20th degree equation to be solved for homework, which he hastily tried to copy down. He succeeded in getting only the first two terms, $x^{20} - 20x^{19}$, before his teacher erased the board completely; however, he did remember that the constant term was $+1$. Can you help our hero by solving the equation?

42. Edges of a Polyhedron

Show that no polyhedron in three-space can have exactly seven edges, while any other integer greater than five is admissible.

43. A Simple Congruence

Show that $63! \equiv 61! \pmod{71}$.

44. The Triangle is Equilateral

If a, b, c are sides of a triangle such that

$$a^2 + b^2 + c^2 = ab + bc + ca,$$

show that the triangle must be equilateral.

45. Simpler Than It Looks

Evaluate the radical

$$\left(\frac{1 \cdot 2 \cdot 4 + 2 \cdot 4 \cdot 8 + 3 \cdot 6 \cdot 12 + \cdots}{1 \cdot 3 \cdot 9 + 2 \cdot 6 \cdot 18 + 3 \cdot 9 \cdot 27 + \cdots}\right)^{1/3}$$

46. Covering a Checkerboard

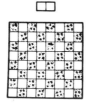

If two squares of opposite colors are removed from anywhere on an 8-by-8 checkerboard, can the remaining portion of the board be completely covered with 1-by-2 dominoes?

47. A Numerical Equality

Show that $(1110)(1111)(1112)(1113) = (1{,}235{,}431)^2 - 1$ in any system of numeration with a base greater than five.

48. Book Publishing

A series of books was published at seven-year intervals. When the seventh book was issued, the sum of the publication years was 13,524. When was the first book published?

49. Three Means

Show geometrically that the geometric mean G of two numbers a and b is the mean proportional between the arithmetic mean A and the harmonic mean H.

50. Beauty Contest

"How about telling me confidentially the secret order of the five beauties to be featured in this year's Annual?" I proposed to the editor. She, of course, refused, but agreed to pass judgment on my guess. "Is it A–B–C–D–E?", I asked.

"You are most skillful at being wrong," she chided. "You not only got each person out of her true position but, furthermore, not one in your ranking followed correctly her immediate predecessor."

"Well, then, is it D–A–E–C–B?", I asked.

"Now you are improving," she encouraged cautiously. "You have two in proper position and you have two following correctly their immediate predecessors."

After a little figuring I then told her the correct order, and she swore me to secrecy. What is the correct order?

51. An Equation Involving Series

Find n if

$$\frac{1^3 + 3^3 + 5^3 + \cdots + (2n-1)^3}{2^3 + 4^3 + 6^3 + \cdots + (2n)^3} = \frac{199}{242}.$$

52. Piled Dominoes

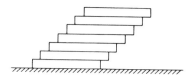

A set of n smooth dominoes 1 inch by 2 inches by $\frac{1}{4}$ inch is piled on a table, one horizontally placed domino in each layer. Find the largest distance that the top domino can be made to overhang the bottom one.

53. Tossing a Die

A die bearing the numbers 0, 1, 2, 3, 4, 5 on its faces is repeatedly thrown until the total of the throws first exceeds 12. What is the most likely total that will be thus obtained?

54. System of Linear Equations

"This system of n linear equations in n unknowns," said the Great Mathematician, "has a curious property."

"Good heavens!" said the Poor Nut, "What is it?"

"Note," said the Great Mathematician, "that the constants are in arithmetic progression."

"It's all so clear when you explain it!" said the Poor Nut. "Do you mean like $6x + 9y = 12$ and $15x + 18y = 21$?"

"Quite so," said the Great Mathematician, pulling out his bassoon.

"Indeed, the system has a unique solution. Can you find it?"
"Good heavens!" cried the Poor Nut, "I am baffled."
Are you?

55. Division by Angle Bisector

Prove that the bisector of an angle of a triangle divides the opposite side into segments proportional to the adjacent sides.

56. Divisibility Probability

Find the probability that if the digits 0, 1, 2, $\cdots$, 9 be placed in random order in the blank spaces of

$$5\_3\,8\,3\_8\_2\_9\,3\,6\_5\_8\_2\,0\,3\_9\_3\_7\,6$$

the resulting number will be divisible by 396.

57. Equation with No Integer Solutions

Prove that the equation $x^2 - 3y^2 = 17$ has no solution in integers.

58. Son of a Mathematics Professor

On the blackboard the mathematics professor wrote a polynomial $f(x)$ with integer coefficients and said, "Today is my son's birthday. When his age A is substituted for x, then $f(A) = A$. You will note also that $f(0) = P$ and that P is a prime number greater than A." How old is the professor's son?

59. Locus in Space

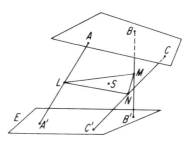

Given a plane E and three noncollinear points A, B, C on the same side of E, the plane ABC not being parallel to E. The points A', B', C' are arbitrary points in E. The points L, M, N are the respective midpoints of AA', BB', CC', and S is the center of mass of the triangle LMN. Find the locus of S as A', B', and C' move in the plane E independently of each other.

60. Weather Analysis

During a period of days, it was observed that when it rained in the afternoon, it had been clear in the morning, and when it rained in the morning, it was clear in the afternoon. It rained on 9 days, and was clear on 6 afternoons and 7 mornings. How long was this period?

61. The Steiner-Lehmus Theorem

Prove that if two internal angle bisectors of a triangle are equal, the triangle is isosceles.

19

62. The Hula Hoop

Consider a vertical girl whose waist is circular, not smooth, and temporarily at rest.   Around the waist rotates a hula hoop of twice its diameter.   Show that after one revolution of the hoop, the point originally in contact with the girl has traveled a distance equal to the perimeter of a square circumscribing the girl's waist.

63. Octasection of a Circle

If k is any real number, show that the lines

$$x^4 + kx^3y - 6x^2y^2 - kxy^3 + y^4 = 0$$

cut the circle $x^2 + y^2 = 1$ into eight equal parts.

64. Arithmetic Progression Devoid of Powers

Find an integral arithmetic progression with an arbitrarily large number of terms such that no term is a perfect rth power for $r = 2, 3, \cdots, n$.

65. Area of Polygon

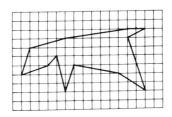

Find the area of the polygon in the figure.

66. A Factorial Equation

Find all the solutions of

$$n!(n - 1)! = m!$$

67. Two Related Triangles

Show that if a, b, c form a triangle, then $\sqrt{a}$, $\sqrt{b}$, $\sqrt{c}$ form a triangle.

68. Maximum Angle in a Circle

Given two points A and B inside a circle, for what point C on the circumference of the circle is the angle ACB the greatest?

69. Determinant of a Magic Square

Let S be the sum of the integer elements of a third-order magic square, and let D be the value of the square considered as a determinant. Show that D/S is an integer.

70. Five-digit Nonsquares

Prove that no perfect square can be written in the decimal scale with just five digits which are distinct, but congruent modulo 2.

71. Dissection of Spherical Surface

How may the total surface of a sphere be divided into the largest possible number of congruent pieces, if each side of each piece is an arc of a great circle less than a quadrant?

72. Trisector of Side of Triangle

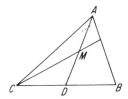

If a line from vertex C of a triangle ABC bisects the median from A, prove that it divides the side AB into the ratio $1:2$.

73. A Factorization

Factor $a^{15} + 1$.

74. A Curious Number

Find a positive number such that $\frac{1}{5}$ of it multiplied by $\frac{1}{7}$ of it equals the number.

75. Two Regular Hexagons

Without using radicals, find the ratio of the areas of the regular hexagons inscribed in and circumscribed about the same circle.

76. General Terms of Series

Find expressions which could be general terms of these series:

(a) 0, 3, 26, 255, 3124, $\cdots$;

(b) 1, 2, 12, 288, 34560, $\cdots$.

77. Heat Flow

Three edges of a square sheet of metal are kept at 0° each, and the fourth edge is kept at 100°. Neglecting surface radiation losses, find the temperature in the middle of the sheet.

78. Does This Make Sense?

If ¼ of 20 is 6, then what is ⅕ of 10?

79. Envelope into Tetrahedrons

CHARLES W. TRIGG
2404 Loring Street
San Diego
California 92109

Dover Publications, Inc.

31 East 2nd Street

Mineola, N.Y. 11501

Can any sealed rectangular envelope, after a single straight cut, be folded into two congruent tetrahedrons?

23

80. A Hammy Cryptarithm

In the following cryptarithm, each letter represents a distinct digit in the decimal scale,

$$7(F\ R\ Y\ H\ A\ M) = 6(H\ A\ M\ F\ R\ Y).$$

Identify the digits.

81. Sheep Buyers

A farmer died and left his two sons a herd of cattle which they sold. The number of dollars received per head was the same as the number of heads. With the proceeds of the sale the sons bought sheep at $10 each and one lamb for less than $10. The sheep and the lamb were divided between the two brothers so that each received the same number of animals. How much should the son who received only sheep pay to the son who received the lamb, in order that the division should be equitable?

82. A Constant Volume

Prove that the volume of a tetrahedron determined by two line segments lying on two skew lines (lines not in the same plane) is unaltered by sliding the segments along their lines but leaving the lengths unaltered.

83. A Repeating Decimal

Compute the first period of the repeating decimal equivalent to $\frac{1}{49}$.

84. A Radical Equation

Solve: $(6x + 28)^{1/3} - (6x - 28)^{1/3} = 2.$

85. The Prize Contest

Professor E. P. B. Umbugio is trying to supplement his meager academic salary by entering soap contests. One such contest requires the contestants to find the number of paths in the following array which spell out the word MATHEMATICIAN:

```
                        M
                      M A M
                    M A T A M
                  M A T H T A M
                M A T H E H T A M
              M A T H E M E H T A M
            M A T H E M A M E H T A M
          M A T H E M A T A M E H T A M
        M A T H E M A T I T A M E H T A M
      M A T H E M A T I C I T A M E H T A M
    M A T H E M A T I C I C I T A M E H T A M
  M A T H E M A T I C I A I C I T A M E H T A M
M A T H E M A T I C I A N A I C I T A M E H T A M
```

Umbugio has counted 1587 paths which originate from one of the first five rows. With the deadline for submitting entries approaching, he is distraught, to say the least. Help the professor out by finding the number of paths with a minimum of computation.

86. A Constant Sum

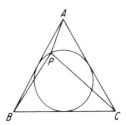

If ABC is an equilateral triangle, and P is any point on the circumference of the inscribed circle, prove that $(PA)^2 + (PB)^2 + (PC)^2$ is constant.

87. Two Incompatible Integers

If x and y are positive integers and $y > 2$, show that $2^x + 1$ is never divisible by $2^y - 1$.

88. A Multiple-choice Question

One and only one of the following pairs of values will not satisfy the equation: $187x - 104y = 41$. Which pair is it?
(1) $x = 3, y = 5$; (2) $x = 107, y = 192$; (3) $x = 211, y = 379$;
(4) $x = 314, y = 565$; (5) $x = 419, y = 753$.

89. Concurrency of Medians

Prove that the medians AA', BB', CC' of a triangle ABC are concurrent.

90. A Surprising Square

In what system of numeration is 11111 a perfect square?

91. Polygon Inscribed in Ellipse?

Show that one cannot inscribe a regular polygon of more than four sides in an ellipse with unequal axes.

92. Telephone Call to Sinkiang

A man is waiting to put through a person-to-person telephone call to Sinkiang. He begins to write the number 0.12345 ⋯ which has n in the nth place of decimals. Being a tidy doodler, he attends promptly to all "carrying figures." Show that he may get his message through before needing to write the digit 8.

93. The Stock Pen

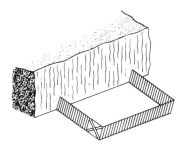

A rancher decided that he needed to fence in $1\frac{1}{4}$ acres for use as a stock pen. A high straight cliff could be used as one side of a rec-

27

tangular area. What dimensions will enable him to fence in the pen at minimum cost?

94. Five Simultaneous Linear Equations

Solve the following set of simultaneous equations:

$$x + y + z + u = 5$$
$$y + z + u + v = 1$$
$$z + u + v + x = 2$$
$$u + v + x + y = 0$$
$$v + x + y + z = 4$$

95. An Almost Universal Theorem

State a theorem about integers which is valid for all integers n, with the exceptions $n = 5, 17,$ and 257.

96. Angle Trisectors

In the triangle ABC, BD and BE are trisectors of angle B, while CD and CE are trisectors of angle C. E is the point closer to side BC. Prove that angles BDE and EDC are equal.

97. Units' Digits of Fibonacci Series

In the Fibonacci series, 1, 1, 2, 3, 5, 8, 13, $\cdots$ wherein the rule of formation is $F_{n+2} = F_n + F_{n+1}$, $F_1 = F_2 = 1$, do the units' digits form a repetitive sequence? That is, one similar to 055055 $\cdots$.

98. Related Cubics

If a, b, c are roots of $x^3 + qx + r = 0$, form the equation whose roots are $(b + c)/a^2$, $(c + a)/b^2$, and $(a + b)/c^2$.

99. A Skeleton Product

The product of three consecutive even integers is $8\,7 * * * * * 8$. Find the integers and supply the missing digits in the product.

100. Inscribed Decagons

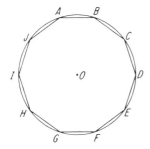 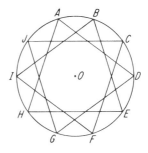

If the circumference of a circle is divided into ten equal parts, the chords joining consecutive points of division form a regular decagon. The chords joining every third division point form an equilateral star decagon. Show that the difference between the sides of these decagons is equal to the radius of the circle.

101. The Handshakers

Every person on earth has shaken a certain number of hands. Prove that the number of persons who have shaken an odd number of hands is even.

102. A Fast Deal

Five cards are drawn at random from a pack of cards which have been numbered consecutively from 1 to 97, and thoroughly shuffled. What is the probability that the numbers on the cards as drawn are in increasing order of magnitude?

103. A Perpendicular Bisector

Prove that the perpendicular bisector of the line joining the feet of two altitudes of a triangle bisects the third side of the triangle.

104. Condition for Divisibility

For what integer a does $x^2 - x + a$ divide $x^{13} + x + 90$?

105. The Farmer's Dilemma

A farmer must buy 100 head of animals with $100. Calves cost $10 each, lambs cost $3 each, and pigs cost 50 cents each. If the farmer buys at least one of each kind of animal, how many of each kind does he buy?

106. A Factored Integer

Find the prime factors of 1,000,027.

107. A Folded Card

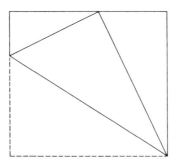

A rectangular card is folded through one corner so that the adjacent corner falls on a side, thus forming three right triangles with areas in arithmetic progression. If the area of the smallest triangle is 3 square inches, what is the area of the largest?

108. A Unique Square

What square is the product of four consecutive odd integers?

109. A Neat Inequality

Show that $n^n > 1 \cdot 3 \cdot 5 \cdot 7 \cdots (2n - 1)$.

110. A Cosine Sum

Evaluate $\cos 5° + \cos 77° + \cos 149° + \cos 221° + \cos 293°$.

111. Quantity Divisible by 9

If $f(x) = x^{10} + x^8 + x^6 + x^4 + x^2 + 1$, show that $f(2i)$ is divisible by 9.

112. Triangular Numbers in Scale of Nine

Show that each member of the infinite series 1, 11, 111, 1111, $\cdots$ is a triangular number in the scale of notation with base nine.

113. Relative Polyhedral Volumes

A regular tetrahedron and a regular octahedron have equal edges. Find the ratio of their volumes without computing the volume of either.

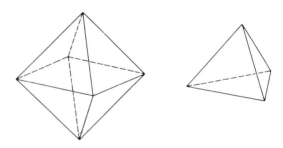

114. The Space Fillers

Show that space can be filled with a tessellation of regular octahedrons and tetrahedrons. (A tessellation is a repetitive space-filling pattern.)

115. Simple Simplification

Reduce 116,690,151/427,863,887 to lowest terms.

116. Sine Sum

Prove that the sum of the sines of a triangle never exceeds $3\sqrt{3}/2$ with equality only when the triangle is equilateral.

117. Two Ferry Boats

Two ferry boats ply back and forth across a river with constant speeds, turning at the banks without loss of time. They leave opposite shores at the same instant, meet for the first time 700 feet from one shore, continue on their ways to the banks, return and meet for the second time 400 feet from the opposite shore. As an oral exercise determine the width of the river.

118. Togetherness at Meals

Albert and Bertha Jones have five children, Christine, Daniel, Elizabeth, Frederick, and Grace. The father decided that he would like to determine a cycle of seating arrangements at their circular dinner table so that each person would sit by every other person exactly once during the cycle of meals. How did he do it?

119. Sum of Digits

Find the sum of the digits appearing in the integers 1, 2, 3, $\cdots$, $(10^n - 1)$.

120. Feeding Three Truck-drivers

Three truck-drivers went into a roadside cafe. One truck-driver purchased four sandwiches, a cup of coffee and ten doughnuts for \$1.69. Another truck-driver purchased three sandwiches, a cup of coffee and seven doughnuts for \$1.26. What did the third truck-driver pay for a sandwich, a cup of coffee and a doughnut?

121. Overlapping Squares

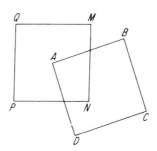

The vertex A of square $ABCD$ is placed so that it coincides with the center of square $MNPQ$ and so that AB trisects MN. If $AB = MN$, find the common area.

122. Solution without Expansion

Without actually expanding solve:

$$(12x - 1)(6x - 1)(4x - 1)(3x - 1) = 5.$$

123. Problem in Primes

A multiplication of a three-place number by a two-place number has the form

$$
\begin{array}{r}
p\ p\ p \\
p\ p \\
\hline
p\ p\ p\ p \\
p\ p\ p\ p \\
\hline
p\ p\ p\ p\ p
\end{array}
$$

The p's are all prime digits, different from unity. Determine their values and show that the solution is unique.

124. Intersecting Great Circles

In general, n great circles on a sphere will intersect in $n(n - 1)$ points. Show how to place the numbers $1, 2, \cdots, n(n - 1)$ on these points in such a way that the sum of the numbers on every great circle is the same. (The plane of a great circle passes through the center of the sphere.)

125. Costly Club

Ten people decided to start a club. If there had been five more in the group, the initial expense to each would have been $100 less. What was the initial cost per person?

126. Binomial Coefficients

What is the largest value of y such that there is a binomial expansion in which the coefficients of y consecutive terms are in the ratio $1:2:3:\cdots:y$? Identify the corresponding expansion and the terms.

127. Quantity Divisible by 8640

Show that for all integer values of x,

$$x^9 - 6x^7 + 9x^5 - 4x^3$$

is divisible by 8640.

128. Dissected Pentagon

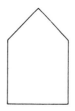

A pentagon consists of a square symmetrically surmounted by an isosceles right triangle. Dissect the pentagon into three pieces which can be reassembled into an isosceles right triangle.

129. Infinite Product

Evaluate the infinite product: $3^{1/3} \cdot 9^{1/9} \cdot 27^{1/27} \cdots (3^n)^{1/3^n}$.

130. Never a Square

Prove that for n a positive integer $n^4 + 2n^3 + 2n^2 + 2n + 1$ is never a perfect square.

131. Chords of a Circle

A circle is divided into n equal parts. Every division point is then connected to every division point m steps away, except that no diameters are drawn. Prove that not more than two of these lines pass through any interior point of the circle.

132. Nine-digit Determinants

The nine positive digits can be arranged into 3-by-3 arrays in 9! ways. Find the sum of the determinants of these arrays.

133. Six Common Points

Find six points common to the graphs of

$$2x^2 + 3xy - 2y^2 - 6x + 3y = 0$$

and

$$3x^2 + 7xy + 2y^2 - 7x + y - 6 = 0.$$

134. A Shuffled Deck

Prove that if the top 26 cards of an ordinary shuffled deck contain more red cards than there are black cards in the bottom 26, then there are in the deck at least three consecutive cards of the same color.

135. Equal Angles

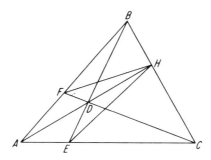

Given an acute-angled triangle ABC and one altitude AH, select any point D on AH, then draw BD and extend it until it intersects

AC in E. Draw CD and extend it until it intersects AB in F. Prove that angle AHE = angle AHF.

136. A Consistent System

For what value of k is the following system consistent?

$$x + y = 1$$
$$kx + y = 2$$
$$x + ky = 3$$

137. Diophantine Equation

Show that the equation $x^2 - y^2 = a^3$ always has integral solutions whenever a is a positive integer.

138. Dissection of Triangle into Two Similar Triangles

Show that any given triangle can be dissected by straight cuts into four pieces which can be arranged to form two triangles similar to the given triangle.

139. The Moving Digits

A number of less than 30 digits begins with the two digits 15 on the left, 15– – – –; and when it is multiplied by 5, the result is merely to move these two digits to the right-hand end, thus, – – – –15. Find the number.

140. Vertex of a Tetrahedron

Show that in every tetrahedron there must be at least one vertex at which each of the face angles is acute.

141. The Lucky Prisoners

A jailer, carrying out the terms of a partial amnesty, unlocked every cell in the prison row. Next he locked every second cell. Then he turned the key in every third cell, locking those cells which were open and opening those cells which were locked. He continued this way, on the nth trip turning the key in every nth cell. Those prisoners whose cells eventually remained open were allowed to go free. Who were the lucky ones?

142. No Point in Common

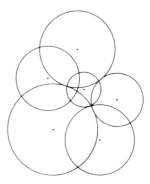

No one of six circular areas in a plane contains the center of another. Show that the six areas have no point in common.

143. Mental Multiplication

Multiply 96 by 104.

144. A Unique Triad

Prove that there is only one set of three distinct positive integers, having no common divisor greater than unity, such that each is a divisor of the sum of the other two.

145. The Enclosed Corner

Across one corner of a rectangular room two 4-foot screens are placed so as to enclose the maximum floor space. Determine their positions.

146. Relatively Prime Integers

If a, b, and c are integers with no factor common to all three, and $1/a + 1/b = 1/c$, show that $(a + b)$, $(a - c)$, and $(b - c)$ are all perfect squares.

147. Diophantine Duo

Solve $a^3 - b^3 - c^3 = 3abc$ and $a^2 = 2(b + c)$ simultaneously in positive integers.

148. Bimedians of a Tetrahedron

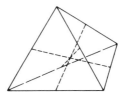

Prove that the lines joining the midpoints of opposite edges of a regular tetrahedron intersect at right angles.

149. Pied Product

When the type for a multiplication of the form $(abc)(bca)(cab)$ was set the digits of the product became pied so that it read 234,235,286. Given that $a > b > c$ and that the correct units' digit is 6, restore the digits of the product to their proper order.

150. A Peculiar Number

If a certain number is reduced by 7 and the remainder is multiplied by 7, the result is the same as when the number is reduced by 11 and the remainder is multiplied by 11. Find the number.

151. Three of a Kind

Prove that at a gathering of any six people, some three of them are either mutual acquaintances or are complete strangers to each other.

152. A Simplification Problem

Simplify $\dfrac{(4 + \sqrt{15})^{3/2} + (4 - \sqrt{15})^{3/2}}{(6 + \sqrt{35})^{3/2} - (6 - \sqrt{35})^{3/2}}$.

153. Product of Three Primes

A certain number is the product of three prime factors, the sum of whose squares is 2331. There are 7560 numbers (including unity) which are less than the number and prime to it. The sum of its divisiors (including unity and the number itself) is 10,560. Find the number.

154. Representation of Rational Number

Prove that any positive rational number can be expressed as a finite sum of distinct terms of the harmonic series, 1, $\frac{1}{2}$, $\frac{1}{3}$, $\frac{1}{4}$, $\cdots$, $1/n$.

155. Condition That a Triangle Is Isosceles

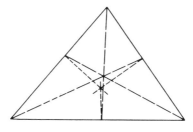

If from the feet of the bisectors of the interior angles of a triangle perpendiculars erected to the respective sides are concurrent, prove that the triangle is isosceles.

156. Classified Integers

Split the integers 1, 2, 3, $\cdots$, 16 into two classes of eight numbers each such that the $C(8, 2) = 28$ sums formed by taking the sums of pairs is the same for both classes.

157. Economical Ballots

A certain physical society is planning a ballot for the election of three officers. There are 3, 4, and 5 candidates for the three offices, respectively. There is a rule in effect (in order to eliminate the ordering of the candidates on the ballot as a possible influence on the election) that for each office, each candidate must appear in each position the same number of times as any other candidate. What is the smallest number of different ballots necessary?

158. Comparison of Ratios

If x and y are positive, which ratio is the greater,

$$(x^2 + y^2) : (x + y) \qquad \text{or} \qquad (x^2 - y^2) : (x - y) ?$$

159. Mixtilinear Triangle

Find the radius of the circle inscribed in the mixtilinear triangle formed by the two legs of a given right triangle ABC and the semicircumference described externally upon the hypotenuse AB.

160. Pandiagonal Heterosquare

We define a pandiagonal heterosquare as a square array of the first n^2 positive integers, so arranged that no two of the rows, columns, and diagonals (broken as well as straight) have the same sum. Is there any n for which these $4n$ sums are consecutive numbers?

161. A Product of 2^{m+1}

Prove that the integer next greater than $(\sqrt{3} + 1)^{2m}$ has a factor 2^{m+1}.

162. In a Nonagon

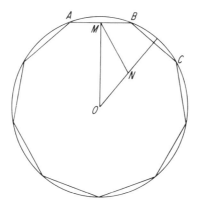

Let AB, BC be two adjacent sides of a regular nonagon inscribed in a circle with center O. Let M be the midpoint of AB and N the midpoint of the radius perpendicular to BC. Show that angle $OMN = 30°$.

163. Spending Money

A father budgeted $6 to distribute equally among his children for spending money at the beach. When two young cousins joined the party and shared in the equal distribution, each child received 25 cents less than had been planned. How many children were in the party?

164. Summing an Infinite Series

Find the sum of the infinite series,

$$1 + 2x + 3x^2 + 4x^3 + \cdots, \ x < 1.$$

165. A Fenced Square Field

A square field is enclosed by a tight board fence of 11-foot boards laid horizontally four boards high. The number of boards in the fence equals the number of acres in the field. What is the size of the field?

166. Triangular Numbers from Odd Squares

Prove that every odd square in the octonary system (scale of eight) ends in 1, and if this be cut off, the remaining part is a triangular number.

167. Inscribed Circles

Which of the following triangles has the larger inscribed circle, one with sides 17, 25, and 26 or one with sides 17, 25, and 28?

168. An Inscribed Dodecagon

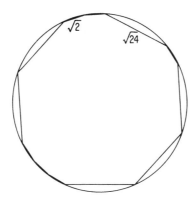

A convex polygon of twelve sides inscribed in a circle has in some order six sides of length $\sqrt{2}$ and six of length $\sqrt{24}$.   What is the radius of the circle?

169. A Susceptible Diophantine Equation

Find a solution in positive integers of $a^3 + b^4 = c^5$.

170. Antifreeze

A 21-quart-capacity car radiator is filled with an 18-percent alcohol solution.   How many quarts must be drained and then replaced by a 90-percent alcohol solution for the resulting solution to contain 42 percent alcohol?

47

171. Maximum-minimum without Calculus

Find the maximum and minimum values of $(x^2 - 2x + 2)/(2x - 2)$ without using the calculus.

172. Tangent Sum Equal to Product

Consider the tangents of $117°$, $118°$ and $125°$.   Prove that their sum is equal to their product.

173. A Faded Document

Reconstruct this "faded document" division in which the illegible digits are represented by asterisks.

```
              * * * *
        * * ) * * * * 0 *
              * *
              ─────
              * * *
              * * 1
              ─────
                * *
                3 *
                ───
```

174. Bisecting Yin and Yang

The monad, or yin and yang, is essentially a circle divided into two equal parts by equal semicircles on opposite sides of a diameter. Bisect each of the equal areas with a single line.

175. Number That Is Factor of Its Reverse

In what system of numeration is 297 a factor of 792?

176. Square Dad

Legally married in California, my neighbor has reached a square age. The product of the digits of his age is his wife's age. The age of their daughter is the sum of the digits of her father's age, and the age of their son is the sum of the digits of his mother's age. How old are they?

177. Tetrahedron through a Straw

Given a flexible, thin-walled cylinder, such as a soda straw, with diameter d. What is the edge e of the largest regular tetrahedron that can be pushed through the straw?

178. A Product of $(a - 1)^2$

Show that $a^{n+1} - n(a - 1) - a$ is divisible by $(a - 1)^2$, where n is a positive integer.

49

179. Determinant of Pascal's Triangle

The arithmetic triangle of binomial coefficients was arranged by Pascal as shown. Prove that the determinant of any square array based on the first row (or column) has unit value.

1	1	1	1	1	1	$\cdots$
1	2	3	4	5	6	$\cdots$
1	3	6	10	15	21	$\cdots$
1	4	10	20	35	56	$\cdots$
1	5	15	35	70	126	$\cdots$
1	6	21	56	126	252	$\cdots$
.	.	.	.	.	.	$\cdots$

180. Rational Coordinates

In the equation $2x^3 + 2y^3 - 3x^2 - 3y^2 + 1 = 0$, if any rational value is assigned to x, show that at least one rational value can be computed for y.

181. A Closed Construction

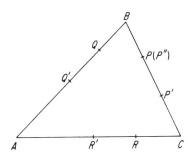

Starting at a point P on the side BC of a triangle ABC, mark Q on AB with $BQ = BP$, R on CA with $AR = AQ$, P' on BC with $CP' = CR$, Q' on AB with $BQ' = BP'$, and so on. Prove that the construction closes, i.e., that $CP = CR'$, and that the six points P, Q, R, P', Q', R' lie on the same circle.

182. Grouped Odd Integers

The consecutive odd integers are grouped as follows: 1; (3, 5); (7, 9, 11); (13, 15, 17, 19); $\cdots$. Find the sum of the numbers in the nth group.

183. The Sixteen-point Sphere

Can the radius of its sixteen-point sphere ever be one-half of the circumradius of a tetrahedron? (The sixteen-point sphere passes through the circumcenters of the faces.)

184. Concurrent Circles

The other three points of intersection of three concurrent circles lie on the same straight line. Prove that their centers and the point of concurrency lie on the same circle.

185. Golf Tournament

A golf pro wishes to arrange a tournament for 16 members of his club. They are to play in foursomes and each member is to play in the same foursome with every other member once and only once. Show how the rounds can be arranged.

186. Square Triangular Numbers

Show that there are infinitely many square triangular numbers.

51

187. System in Three Variables

Solve the system: $x + y + z = 6$, $xy + yz + zx = 11$, $xyz = 6$.

188. Imbedded Polyhedron

The vertices and edges of a certain polyhedron, from which the inside and the faces have been removed, may be imbedded in the plane, allowing stretching. Sketch the original polyhedron if the plane imbedding is as shown.

189. Baseball Team Standings

Following is the won-lost record of the baseball teams in the National League on July 14, 1965:

	W	L		W	L
Chicago	41	46	New York	29	56
Cincinnati	49	36	Philadelphia	45	39
Houston	39	45	Pittsburgh	44	43
Los Angeles	51	38	St. Louis	41	45
Milwaukee	42	40	San Francisco	45	38

Arrange the teams in correct descending order of "percentage" won without calculating the "percentages."

190. A Sweet Purchase

A housewife purchased some sugar for \$2.16.   Had the sugar cost 1 cent a pound less, she would have received 3 pounds more for the same expenditure.   How many pounds of sugar did she buy?

191. Intersections of Diagonals

Find the number of intersections of the diagonals of a convex polygon of n sides.

192. Two Vanishing Triads

Prove that if the sum of the members of each of two number triads is zero, then the sums of the cubes of the members of the triads are in the same ratio as the products of the members.

193. Condition for Factorability

If a and b are prime to 3 and $a + b$ is of the form $3k$, then show that $x^a + x^b + 1$ is factorable.

194. Parallel Resistances

The equivalent electrical resistance z of two resistances x and y connected in parallel is given by the relationship

$$1/z = 1/x + 1/y, \qquad x, y, z > 0.$$

Determine the solution in positive integers.

195. Minimum Bisector

What is the curve of minimum length which bisects the area of an equilateral triangle?

196. A Peculiar Square

Find a nine-digit integer of the form $a_1a_2a_3b_1b_2b_3a_1a_2a_3$ which is the product of the squares of four distinct primes, $a_1 \neq 0$, $b_1b_2b_3 = 2(a_1a_2a_3)$.

197. A Series of Tests

Professor Tester gave marks based upon an average of a series of tests. As John came into the last test, he realized that he would have to make a 97 in order to average 90 for the course. On the other hand, if he was as low as 73, he would still be able to average 87. How many tests were in Professor Tester's series?

198. Coinciding Points in a Quadrilateral

Show that the midpoint of the line segment joining the midpoints of the diagonals of a quadrilateral coincides with the point of intersection of the line segments joining the midpoints of the opposite sides of the quadrilateral.

199. Fractions in Lowest Terms

Show that in the sequence $1/n$, $2/n$, $3/n$, $\cdots$, $(n - 1)/n$, where n is a positive integer greater than 2, an even number of the members are fractions in lowest terms.

200. The Tea Set

A silver tea set in a dealer's window had the following cost marks and retail prices:

Sugar bowl........$HKHC$ $6.72   Creamer.....$HCKH$ $6.00
Tray............$AMSL$ 50.16   Teapot......$SIAB$ 91.08
Tongs...........$HBLT$ 1.72   Spoons......$HMIT$ 10.52
            Complete set................$BLCSK$ $166.20

If the markup was the same percent of the cost in each case, break the cost-mark code.

201. Segments Determining an Equilateral Triangle

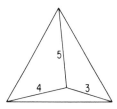

Three segments, 3, 4, and 5 inches long, one from each vertex of an equilateral triangle, meet at an interior point P. How long is the side of the triangle?

202. An Invariant Remainder

What number when divided into 1108, 1453, 1844, and 2281, always leaves the same remainder?

203. Nine Non-Zero Digits

Find a permutation of the nine non-zero digits with a square root of the form $ababc$ where $ab = c^3$.

204. Dissection for Coincidence

Consider two congruent triangles that can be brought into coincidence only by a rotation of one of them through a third dimension. How, by cutting the triangles, could coincidence be effected by motion in a plane only?

205. Deflating Umbugio

Professor E. P. B. Umbugio has recently been strutting around because he hit upon the solution of the fourth-degree equation which results when the radicals are eliminated from the equation

$$x = (x - 1/x)^{1/2} + (1 - 1/x)^{1/2}.$$

Deflate the professor by solving this equation using nothing higher than quadratic equations.

206. Males with Common Characteristics

If 70 percent of the adult males of a community have brown eyes, 75 percent have dark hair, 85 percent are over 5 feet 8 inches tall, and 90 percent weigh more than 140 pounds, what percent at least have all four characteristics?

207. An Area of Constant Width

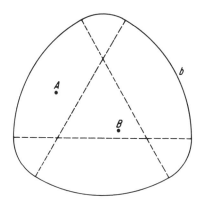

Given two points A and B within a region of constant width 1. Show that there is a path from A to B which touches the boundary b of the region and whose length ≤ 1. [A curve of constant width is a curve such that the distance between parallel tangents (support lines) is constant.]

208. Comparing Radicals

Which is the larger, $\sqrt[8]{8!}$ or $\sqrt[9]{9!}$?

209. Fibonacci Tetrahedron

Find the volume of the tetrahedron with vertices (F_n, F_{n+1}, F_{n+2}), $(F_{n+3}, F_{n+4}, F_{n+5})$, $(F_{n+6}, F_{n+7}, F_{n+8})$, and $(F_{n+9}, F_{n+10}, F_{n+11})$, where F_i is the ith Fibonacci number in the sequence, 1, 1, 2, 3, 5, 8, $\cdots$.

210. The Regular Octahedron

A regular octahedron, edge e, is cut by a plane parallel to one of its faces. Find (a) the perimeter, and (b) the area of the section.

211. Never an Integer

Prove that $1 + \frac{1}{2} + \frac{1}{3} + \cdots + 1/n$ is never an integer for any $n > 1$.

212. Three Consecutive Odd Integers

Show that there are no three consecutive odd integers such that each is the sum of two squares greater than zero.

213. Vanishing Vector Sum

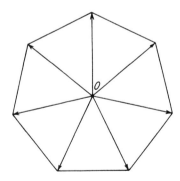

Prove that the sum of the vectors from the center O of a regular n-gon to its vertices is zero.

214. The Sliding Ellipse

Two fixed lines are perpendicular to each other. An ellipse moves so that it is always tangent to both lines. Find the locus of its center.

215. Superposed Radical

Evaluate $\sqrt[3]{11 + 4\sqrt[3]{14 + 10\sqrt[3]{17 + 18\sqrt[3]{(\cdots)}}}}$.

216. Fibonacci Pythagorean Triangles

Find Pythagorean triangles whose sides are Fibonacci numbers.

217. Hole in Sphere

The axis of a cylindrical hole 10 inches long coincides with a diameter of a sphere through which it is drilled. What is the volume of the material remaining?

218. The Commuter

If a man walks to work and rides back home it takes him an hour and a half. When he rides both ways it takes 30 minutes. How long would it take him to make the round trip by walking?

219. Odd Base of Notation

Show that any integer in a system of notation with an odd base is odd, if and only if it has an odd number of odd digits.

220. Journey on a Dodecahedron

Suppose that the vertices of a polyhedron represent places that we want to visit, while the edges represent the only possible routes. Hamilton considered the problem of visiting all the places, without repetition, on a single journey. [See, for example, W. W. R. Ball and H. S. M. Coxeter, *Mathematical Recreations and Essays*, The Macmillan Company, New York, (1956), p. 262.] This is easily solved for the pentagonal dodecahedron. Prove that it cannot be done for the rhombic dodecahedron.

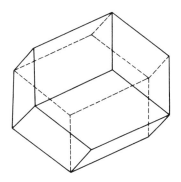

221. Rhombic Dodecahedrons

Show that space can be filled with rhombic dodecahedrons.

222. A Fibonacci Relationship

For the Fibonacci sequence $\{F_n\}$, where $F_1 = F_2 = 1$, $F_n = F_{n-1} + F_{n-2}$, $n \geq 3$, show that every fifth number in the sequence is divisible by 5.

223. Cryptic Multiplication

After making an ordinary arithmetic multiplication, a
man replaced every even digit with E and every odd digit
with O. He obtained the arrangement on the right.
What was the multiplication?

$$
\begin{array}{r}
O\ E\ E \\
E\ E \\
\hline
E\ O\ E\ E \\
E\ O \\
\hline
O\ O\ E\ E
\end{array}
$$

224. Creased Rectangle

Two opposite vertices of an x-by-y rectangle are brought into
coincidence, and the rectangle is flattened out to form a crease.
Find the length of the crease.

225. A Man's Birthdate

In 1937 a man stated that he was x years old in the year x^2. He
added, "If the number of my years be added to the number of my
month, the result equals the square of the day of the month on which
I was born." When was he born?

226. A Fractional Equation

Solve the equation

$$(x - a)/b + (x - b)/a = b/(x - a) + a/(x - b).$$

61

227. A Particular Isosceles Triangle

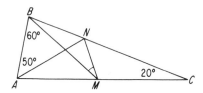

An isosceles triangle ABC has a vertex angle $C = 20°$. Points M and N are so taken on AC and BC that angle $ABM = 60°$ and angle $BAN = 50°$. Without resorting to trigonometry, prove that angle $BMN = 30°$.

228. Tree Leaves

If there are more trees than there are leaves on any one tree, then there exist at least two trees with the same number of leaves. True or false?

229. A Diophantine Cubic

Solve $x^3 + 1 = y^2$ in integers.

230. Countries on a Sphere

Suppose that the surface of a sphere is divided into triangular "countries," where *triangular* means that each country touches exactly three others. A vertex of the graph formed by the boundary lines of the countries is called *even* or *odd* depending on whether an even or an odd number of boundary lines runs into it. Is there such a triangulation having exactly two odd vertices in which these vertices are adjacent?

231. Henry's Trip

Henry started a trip into the country between 8 A.M. and 9 A.M. when the hands of the clock were together. He arrived at his destination between 2 P.M. and 3 P.M. when the hands of the clock were exactly 180° apart. How long did he travel?

232. A Power Series

Express $1/(1 + x)(1 + x^2)(1 + x^4)(1 + x^8)$ as a power series.

233. Parallels in a Triangle

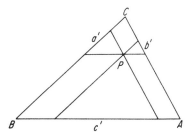

Through any interior point P in triangle ABC, lines are drawn parallel to the three sides of the triangle, dividing these sides into three segments each. If the middle segments of sides a, b, and c are denoted by a', b', and c', respectively, show that

$$a'/a + b'/b + c'/c = 1.$$

234. A Partitioning Problem

Partition 316 into two parts so that one part is divisible by 13 and the other part by 11.

235. Fair Fares

Find the digits in the scale of six which are uniquely represented by the letters in $F\,A\,R\,E\,S = (F\,E\,E)^2$.

236. How Old Is Willie?

"Did your teacher give you that problem?" I asked. "It looks rather tedious."

"No," said Willie, "I made it up. It's a polynomial equation with my age as a root. That is, x stands for my age my last birthday."

"Well, then," I remarked, "It shouldn't be so hard to work out —integer coefficients, integral root. Suppose I try $x = 7 \cdots$. No, that gives 77."

"Do I look only seven years old?" demanded Willie.

"Well, let me try a larger integer $\cdots$. No, that gives 85, not zero."

"Oh, stop kidding!" said Willie, looking over my shoulder. "You know I'm older than that."

How old is Willie?

237. Accelerating Particle

A particle moves in a straight line starting from rest and finishing at rest, and covers unit distance in unit time. Prove that at some point its acceleration has a magnitude of at least 4 units. It is assumed that v and a are continuous functions of t.

238. Stamps for Buck

A boy sent to buy $1 worth of stamps asked for some 2-cent stamps, ten times as many 1-cent stamps and the rest in 5-cent stamps. How many of each did he buy?

239. A Twenty Question Game

As a variation of the game of *Twenty Questions*, suppose I think of a number which you are to determine by asking me not more than twenty questions, each of which can be answered by only "yes" or "no." What is the largest number that I should be permitted to choose so that you may determine it in twenty questions?

240. Inscribed Spheres

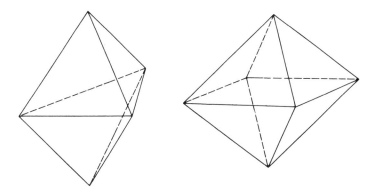

If the faces of a hexahedron are equilateral triangles congruent to the faces of a regular octahedron, find the ratio of the radii of the inscribed spheres.

241. Disks from a Disk

One disk 20 inches in diameter and one 10 inches in diameter are cut from a disk of plywood 30 inches in diameter. What is the largest disk that can be cut from the remainder of the plywood?

242. Sum of Squares of Binomial Coefficients

Find the sum of the squares of the coefficients in the expansion of $(a + b)^n$.

243. Christmas Cryptarithm

The holiday greeting, *A MERRY XMAS TO ALL*, is a cryptarithm in which each of the letters is the unique representation of a digit in the decimal scale, and each word is a square integer. Find the numerical interpretation if in each word the sum of the digits is also a square.

244. Centers of Gravity

Find the center of gravity of (a) a semicircular wire, and (b) a semicircular area.

245. Two Equal Triads

No one of the three positive numbers x, y, z is less than the smallest of the three positive numbers a, b, c, nor is any one of x, y, z greater than the largest of a, b, c. If $x + y + z = a + b + c$ and $xyz = abc$, show that, in some order, x, y, z are equal to a, b, c.

246. An Irrational Sum

Prove that $\sum_{n=1}^{\infty} 6^{(2-3n-n^2)/2}$ is irrational.

247. Examination of Six Students

Suppose six students be standing an examination in a row of seats with an aisle at each end. If they finish in random order, what is the probability that a student will have to pass over one or more other students in order to reach an aisle?

248. Construction by Compasses Alone

By use of compasses alone divide the circumference of a circle into four equal parts.

249. The Bonus Fund

It was planned to distribute fifty dollars of a bonus fund to each employee, but the last man would have gotten only forty-five dollars. In order to effect an equitable distribution, forty-five dollars was given to each person, and ninety-five dollars was kept in the fund for the following year. How much money was in the fund to begin with?

250. The Court Mathematician's Salary

The court mathematician once received his salary for a year's service all at one time, and all in silver "dollars," which he proceeded to arrange in nine unequal piles, making a magic square. The king looked, and admired, but complained that there was not a single prime number in any of the piles. "If I had but nine coins more," said the mathematician, "I could add one coin to each pile and make a magic square with every number prime." They investigated and found that this was indeed true. The king was about to give him nine "dollars" more, when the court jester said, "Wait!" Then the jester subtracted one coin from each pile instead, and they found in this case also a magic square with every element a prime number. The jester kept the nine "dollars." How much salary must the mathematician have been receiving?

251. Packing Cylinders

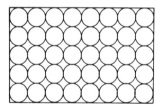

Forty cylinders, with 1-inch diameters and equal heights, are packed snugly in five rows of eight each in a box so that they may be transported without rattling. How many must be taken from the box in order to repack it with forty-one of the same sized cylinders? Will they now rattle?

252. Leg of a Pythagorean Triangle

Show that the length of one leg of a Pythagorean triangle must be a multiple of 3.

253. Powers of Two

Sum the series, $2 + 2^2 + 2^3 + 2^4 + \cdots + 2^n$.

254. Colorful Square Arrays

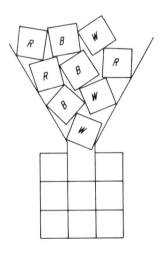

In how many distinct patterns can 9 congruent squares—3 red, 3 white, and 3 blue—be arranged in a square array so that all three colors occur in every column and row?

255. Bisected Parallelogram

Points E, F are taken on the sides BC and AD, respectively, of parallelogram $ABCD$. Now AE intersects BF at G and ED intersects CF at H. Prove that GH prolonged bisects the parallelogram.

256. Squares of Reverse Integers

A square number in the scale of six is composed of the five positive digits. If its units' digit be transferred to the front of the number, the digits of the square root are reversed. Find the number.

257. A Questionable Sum

If a and b are integers, can $a/b + b/a$ ever be an integer?

258. Is the Square Fault-Free?

Can eighteen 2-inch by 1-inch dominoes be assembled into a fault-free square? That is, assembled into a square in which no straight line formed by the edges of the dominoes joins opposite sides?

259. Relatively Prime Numbers

In what systems of notation are 35 and 58 relatively prime?

260. Five Consecutive Integers

Are there any five consecutive positive integers such that the sum of the first four, each raised to the fourth power, equals the fifth raised to the fourth power?

261. Polygonal Path in a Lattice

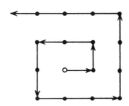

Given an N by N point lattice with $N > 2$, show that it is possible to draw a polygonal path passing through all the N^2 lattice points and consisting of $2N - 2$ segments. (The procedure in the figure requires $2N - 1$ segments.)

262. Wanted—Integer Solutions

Find all solutions in integers of the equation

$$y^2 + y = x^4 + x^3 + x^2 + x.$$

263. A Composite Number

p_1 and p_2 are consecutive odd primes, so $p_1 + p_2 = 2q$. Show that q is composite.

264. Four Simultaneous Linear Equations

Solve the following system of equations:

$$x + 7y + 3v + 5u = 16 \tag{1}$$

$$8x + 4y + 6v + 2u = -16 \tag{2}$$

$$2x + 6y + 4v + 8u = 16 \tag{3}$$

$$5x + 3y + 7v + u = -16 \tag{4}$$

265. Property of a Quadrangle

A quadrangle with area Q is divided by its diagonals into four triangles with areas A, B, C, and D. Show that

$$(A)(B)(C)(D) = (A + B)^2(B + C)^2(C + D)^2(D + A)^2/Q^4.$$

266. When Is the Division Exact?

For what positive integral values of n does $2n + 1$ divide $n^4 + n^2$?

267. A Doubtful Equation

The statement "3 4 2 = 9 7" can be made true by inserting a few algebraic signs, thus: $(-3 + 4) 2 = 9 - 7$. Can its verity be established without inserting any signs?

268. Dissected Dodecagon

(a) Dissect a regular dodecagon into squares and equilateral triangles.

(b) Let P_1, $P_2 \cdots$, P_{12} be the consecutive vertices of the regular dodecagon. Explain how the diagonals P_1P_9, P_2P_{11}, and P_4P_{12} intersect.

269. No Real Roots

Show that

$$1 + x + x^2/2! + x^3/3! + \cdots + x^{2n}/(2n)! = 0$$

has no real roots.

270. Impossible Cube

Show that there is no scale of notation in which the three-digit number $aaa = a^3$.

QUICKIE
SOLUTIONS

The most elegant solution which this writer has found for each of the challenge problems is given on the following pages together with the source and author in accordance with the statement in the Preface. Readers are invited to submit any more elegant solutions which they may devise.

The solutions are numbered consecutively to correspond to the numbers of the challenge problems. A Q is prefaced to the number of the quickie solution to distinguish it from the challenge problem. Numbers of the solutions printed on a page appear at the upper outside corner of the page.

Q 1. The Careless Mailing Clerk

If nine letters are in the correct envelopes, the tenth must be also, so the probability is zero.

—*M.M.*, 33 (March, 1950), 210.

Q 2. The Pythagorean Theorem

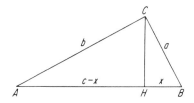

In the right triangle ABC, draw CH, the altitude to the hypotenuse Then in the figure, triangles ACB, AHC, and CHB are similar, whereupon

$$x:a = a:c \quad \text{and} \quad (c - x):b = b:c,$$

so
$$a^2 = cx \quad \text{and} \quad b^2 = c^2 - cx$$

Adding, $a^2 + b^2 = c^2$.

For this and 365 other ways of proving the Pythagorean theorem, see Elisha S. Loomis, *The Pythagorean Proposition*, Edwards Brothers, Ann Arbor, Michigan (1940).

Q 3. Four Equations in Four Unknowns

By inspection $(1, 2, 3, 4)$ is a solution of the first and fourth equations and satisfies the second and third equations.   Since the equations are symmetrical in x, y, z, w, the other 23 permutations of 1, 2, 3, 4 are solutions also.   But these are all the solutions, since the product of the degrees of the equations is 4!

—*M.M.*, 23 (March, 1950), 211.

Q 4. A Test with Zero Score

The number of answers in each category is inversely proportional to the value, so there were $[5/(5 + 8)](26)$ or 10 correct answers.

—*M.M.*, 31 (March, 1958), 237.

Q 5. Ptolemy's Theorem

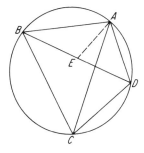

In the inscribed quadrilateral $ABCD$ draw AE making angle BAE = angle CAD.   Then triangles BEA and CDA are similar, and so are triangles AED and ABC.   Hence, $AC:AB = CD:BE$ and $AC:AD = BC:ED$.   Consequently, $(AC)(BE) = (AB)(CD)$

and $(AC)(ED) = (AD)(BC)$. Adding, and noting that $BE + ED = BD$, we have

$$(AC)(BD) = (AB)(CD) + (AD)(BC).$$

This is Ptolemy's theorem.

—Roy MacKay, *S.S.M.*, 35 (March, 1935), 314.

If the inscribed quadrilateral is a rectangle, the Pythagorean theorem follows at once.

Q 6. Easy Factoring

We have $x^9 + y^9 = (x + y)(x^8 - x^7y + x^6y^2 - \cdots + y^8)$

$$= (x^3 + y^3)(x^6 - x^3y^3 + y^6)$$

$$= (x + y)(x^2 - xy + y^2)(x^6 - x^3y^3 + y^6).$$

Hence,

$$x^8 - x^7y + x^6y^2 + \cdots + y^8 = (x^2 - xy + y^2)(x^6 - x^3y^3 + y^6).$$

—Anice Seybold, *M.M.*, 34 (November, 1961), 434.

Q 7. Mental Computation

Since $(10a + 5)^2 = 100a^2 + 100a + 25 = a(a + 1)(100) + 25$, then $(85)^2 = (8)(9)(100) + 25 = 7225$.

—*M.M.*, 24 (May, 1951), 273.

Q 8. A Quartic Equation

The equation $x^4 - 5x^3 - 4x^2 - 7x + 4 = 0$ may be written in the form $(x^2 - 2)^2 = 5x^3 + 7x$. For every negative x, the left-hand

79

member is non-negative and the right-hand member is negative, so no negative x can satisfy the original equation.

—R. E. Horton, *M.M.*, 24 (November, 1950), 114.

Q 9. A Million on Each Side

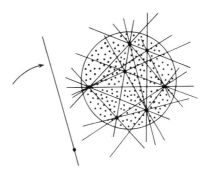

Consider all the lines determined by pairs of points in the array. Pick a new point belonging to none of these lines and outside the circle. Consider a line through this new point and to the left of the array of points. As this line is rotated about the new point and toward the right of the array, it passes exactly one point of the array at a time. Hence, rotate this line until it has passed through exactly one million points, at which location it becomes the line sought.

—Herbert Wills, *M.M.*, 37 (May, 1964), 206.

Q 10. An Infinitude of Primes

Suppose that there is a largest prime p. Consider the quantity which is one greater than the product of all the primes less than or equal to p, that is,

$$Q = 2 \cdot 3 \cdot 5 \cdot 7 \cdots p + 1.$$

Now Q is not divisible by any of the primes in the product (since there always is a remainder of 1). Hence Q is either prime or if it is composite it must be the product of primes all greater than p. In either event, there is a prime greater than p. Therefore there is no largest prime.

—Euclid, *Elements*, Book IX, Proposition 20.

Even though it is well known, this beautiful classic proof belongs in any collection of solutions purporting to be elegant.

Q 11. Complex Numbers

Since $i = \sqrt{-1}$, $i^8 = 1$, so

$$\frac{27 + 8i}{3 + 2i^3} = \frac{27 + 8i^9}{3 + 2i^3} = 9 - 6i^3 + 4i^6 = 5 + 6i.$$

—J. M. Howell, *M.M.*, 26 (May, 1953), 287.

Q 12. Overlapping Areas

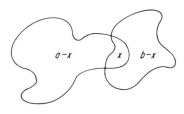

In general, if any two areas a and b have a common area x, then the nonoverlapping portions are $a - x$ and $b - x$.   The difference of the areas of the nonoverlapping portions is $|a - b|$.   In this problem, the difference is $\pi(20)^2 - \pi(15)^2$ or 175π.

—*M.M.*, 26 (May, 1953), 287.

Q 13. A Tennis Tournament

Each match has one loser, each loser loses only once, so there are $n - 1$ losers, hence $n - 1$ matches.

—Fred Marer, *M.M.*, 23 (May, 1950), 278.

Q 14. Summing a Factorial Series

We have

$$1(1!) + 2(2!) + 3(3!) + \cdots + (n-1)[(n-1)!] + n(n!)$$
$$= 2(1!) + 3(2!) + 4(3!) + \cdots + n[(n-1)!] + (n+1)(n!)$$
$$-\;1!\;-\;\;2!\;-\;\;\;3!\;-\cdots-\;\;\;(n-1)!\;-\;\;\;\;\;\;\;\;\;n!$$
$$= (n+1)! - 1.$$

Q 15. Intersecting Cylinders

Consider any cross section of the common volume parallel to the square midsection.   This will be a square whose inscribed circle will be the cross section of the inscribed sphere of the common volume. Hence the common volume is to the volume of the sphere as the area of the square is to the area of its inscribed circle.   Then

$$V = (4\pi r^3/3)(4/\pi) \text{ or } 16\,r^3/3.$$

—Leo Moser, *M.M.*, 25 (May, 1952), 290.

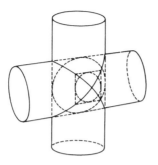

It follows that the common space can be subdivided into infinitesimal pyramids with vertices at the intersection of the axes and bases along the elements of the cylinders. These pyramids have unit altitude. Hence the area of the surface of the common volume is 16.

—J. H. Butchart, M.M., 26 (September, 1952), 54.

Q 16. Representations of an Integer

Consider n 1's in a row with spaces between them. There is clearly a one-to-one correspondence between expressions for n as a sum and ways of disposing of the $n - 1$ spaces by entering plus signs or leaving the spaces blank. This gives $n - 1$ tasks to perform and two ways of handling each one. Thus the total number of expressions for n as a sum is 2^{n-1}.

—William Moser, P.M.E.J., 1 (November, 1951), 186.

Q 17. Quartic with a Rational Root

In any polynomial $f(x)$, $f(1) =$ the sum of the coefficients. If this sum is zero, then $(x - 1)$ is a factor of $f(x)$. Since $1 - 2 +$

$3 + 4 - 6 = 0$, it follows that $x = 1$ is a root of the equation regardless of the order in which the gaps are filled.

Q 18. Resistance of a Cube

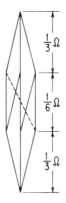

Consider the edges of the cube to be hinged at the vertices.   Upon lifting the cube by a vertex, the edges will fall into a series of three parallel, six parallel, and three parallel edges terminating at the opposite corner.   Hence the overall resistance is $\frac{1}{3} + \frac{1}{6} + \frac{1}{3}$ or $\frac{5}{6}$ ohms.

Q 19. A Facetious Division

Clearly

$$J\,O\,K\,E = \frac{A\,H\,H\,A\,A\,H}{H\,A} = 100 + \frac{A\,H\,(10{,}001)}{H\,A}.$$

Now $10{,}001 = (73)\,(137)$.   There is no combination of distinct digits such that $H\,A$ divides $A\,H$.   Therefore $H\,A$ divides $10{,}001$ and $H\,A = 73$.   Consequently, $377{,}337/73 = 5169$.

Q 20. The Flower Salesman

Since y is an integer < 2, $y = 1$. Then, dealing with the price per flower in cents,

$$\frac{100}{x} - \frac{200}{x + 10} = \frac{80}{12} \qquad \text{or} \qquad x^2 + 25x - 150 = 0$$

of which the positive root is $x = 5$, the number of flowers originally purchased.

Q 21. Relative Polygonal Areas

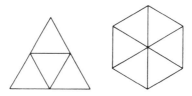

The sides of the triangle and the hexagon are in the ratio 2:1. Hence the triangle can be dissected into four equilateral triangles and the hexagon into six equilateral triangles, all congruent. So the areas are in the ratio 2:3.

—M.M., 34 (May, 1961), 308.

Q 22. The Inverted Cups

If n is *even*, and we use n moves each leaving a different cup unchanged, then finally each cup will have been inverted $n - 1$ times, so the entire set will be inverted.

If n is *odd*, represent each upright cup by $+1$ and each inverted cup by -1. Then we start with the product of all the representations as $+1$ and desire to get -1. However, each move which inverts $n - 1$ cups, an even number, leaves the product as $+1$.

—E. P. Starke.

Q 23. The End of the World

That $x - y$ is a divisor of $x^n - y^n$ for $n = 0, 1, 2, \cdots$, is the only principle required. Let the professor's number be $F(n)$. Then since $2141 - 1863 = 1770 - 1492 = 278$, $F(n)$ is always divisible by 278. Similarly, $2141 - 1770 = 1863 - 1492 = 371$, which is relatively prime to 278. So $F(n)$ is always divisible by $(278)(371) = (53)(1946)$, and hence, of course, by 1946 itself.

—E. P. Starke, *A.M.M.*, 54 (January, 1947), 43.

Q 24. Six Distinct Integers

If b is the unit repetend of a, then all the digits of ab are 9's. Now $99 = (9)(11)$, $999 = (3)(9)(37)$, $9999 = (9)(11)(101)$, $99999 = (9)(41)(271)$, and $999999 = (3)(7)(9)(11)(13)(37)$. The last set is the smallest set of six distinct integers having the stated property. In fact,

$$\tfrac{1}{3} = 0.333333\cdots \qquad\qquad \tfrac{1}{11} = 0.090909\cdots$$

$$\tfrac{1}{7} = 0.142857\cdots \qquad\qquad \tfrac{1}{13} = 0.076923\cdots$$

$$\tfrac{1}{9} = 0.111111\cdots \qquad\qquad \tfrac{1}{37} = 0.027027\cdots$$

—*M.M.*, 35 (November, 1962), 311.

Q 25. Length of a Helix

Roll the cylindrical surface and wire onto a plane.   The element (9 inches), the repeated circumference (10·4 inches), and the wire (L) now form a right triangle.   Hence, $L = (81 + 1600)^{1/2}$ or 41 inches of wire.

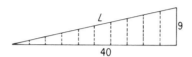

It follows that unless a coil of wire is wrapped around a cylindrical surface so as to cut the elements of the cylinder at a constant angle, there is danger that it may become loosened with time.

—*M.M.*, 23 (May, 1950), 278.

Q 26. A Minimum Problem

The fraction may be written in the form

$$(p + 1 + 1/p)(q + 1 + 1/q)(r + 1 + 1/r)(s + 1 + 1/s).$$

Now the sum of a positive number and its reciprocal is ≥ 2.   Hence each factor of the product is ≥ 3 and the product is ≥ 81.

—R. L. Moenter, *S.S.M.*, 54 (November, 1954), 667.

It follows immediately that

$$\frac{(a_1^2 + a_1 + 1)(a_2^2 + a_2 + 1) \cdots (a_n^2 + a_n + 1)}{a_1 a_2 \cdots a_n} \geq 3^n$$

for positive a_i.

Q 27. Digits of a Square Number

Examination of a table of squares in the decimal system shows that:

1. A square can end only in 0, 1, 4, 5, 6 or 9.

2. The tens' digit of a square number is even unless the units' digit is 6, in which event it is odd.

Thus the digits cannot all be the same unless they are all 4's. (A square consisting only of zeros is obviously meaningless.) But $\cdots444 = 4(\cdots111)$ and since the tens' digit inside the parentheses is odd, no square integer in the decimal system can consist of digits all alike.

N	N^2
. .	. . .
10	100
11	121
12	144
13	169
14	196
15	225
16	256
17	289
18	324
19	361
. .	. . .

Q 28. Interlocking Committees

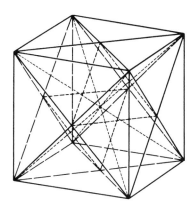

Take the directors as the vertices, center and centers of the faces of a cube. Sixteen of the committees are the diagonals of the cube

and its faces, giving four committees per director except for the face centers, which are only on two committees.   But these face centers form an octahedron, and we can form the additional committees from alternate triangles on the octahedron.

—J. Evernden and R. Spira, *A.M.M.*, 69 (November, 1962), 921.

Q 29. The Jigsaw Puzzle

Since the number of pieces is originally n and finally is 1, and since every move reduces the number of pieces by 1, the total number of moves required is $n - 1$, and clearly is independent of the procedure.   So long, of course, as we never separate pieces already joined.

—Leo Moser, *M.M.*, 26 (January, 1953), 169.

Q 30. A Remainder Problem

Since $f(x) = x^4 + x^3 + x^2 + x + 1$, $(x - 1) \cdot f(x) = x^5 - 1$. Then $f(x^5) = (x^{20} - 1) + (x^{15} - 1) + (x^{10} - 1) + (x^5 - 1) + 4 + 1$. $x^5 - 1$ and hence $f(x)$ is a factor of each of the quantities in parentheses, so $f(x^5) = [$a multiple of $f(x)] + 5$.

—Norman Anning, *S.S.M.*, 54 (October, 1954), 576.

Q 31. A Skeleton Division

If we denote the divisor by d, we have $8d < 1000$, so $d < 125$. Then since $7d < 900$, it follows from the first subtraction that the first digit of the quotient is 8.   Consequently the quotient is 80,809.

Since $80,809d > 10,000,000$, we have $d > 123$. Hence $d = 124$, and the reconstructed division is

$$
\begin{array}{r}
80809 \\
124\overline{)10020316} \\
992 \quad\;\; \\
\overline{1003\;\;\;} \\
992\;\; \\
\overline{1116} \\
1116 \\
\end{array}
$$

—W. B. Carver, *A.M.M.*, 61 (December, 1954), 712.

Q 32. Triangles in a Circle

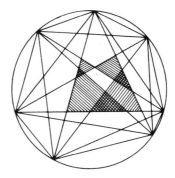

Every set of six points on the circumference of the circle can be paired in one and only one way such that the three lines joining pairs will form an admissible triangle. Conversely, every admissible triangle has sides leading to six points on the circumference. Hence the number of admissible triangles is $C(n, 6)$ or $n!/6!(n - 6)!$

—Leo Moser, *M.M.*, 26 (March, 1953), 226.

The figure shows the case $n = 7$.

Q 33. Easy Multiplication

Since $125 = 1000/8$, $(5,746,320,819)(125) = 5,746,320,819,000/8 = 718,290,102,375$.

—*M.M.*, 25 (May, 1952), 289.

Q 34. A Repetitive Series

Subtract $\frac{1}{2}(7 - 4)$ or 1.5 from each term of

$$- 4 + 7 - 4 + 7 - 4 + 7 - \cdots$$

to obtain

$$- 5.5 + 5.5 - 5.5 + 5.5 - 5.5 + 5.5 - \cdots.$$

Thus the nth term of the series is $1.5 + 5.5(-1)^n$.

Q 35. Radical Simplification

Let $\sqrt[3]{2 + \sqrt{5}} = a$, $\sqrt[3]{2 - \sqrt{5}} = b$, and $a + b = x$.

Then

$$x^3 = a^3 + 3a^2b + 3ab^2 + b^3 = a^3 + b^3 + 3ab(a + b)$$

$$= 4 + (3)(\sqrt[3]{-1})x.$$

So $x^3 + 3x - 4 = 0$ and the only real root of this equation is 1.

—Claire Adler, A.M.M., 59 (May, 1952), 328.

Otherwise: $[(1 + \sqrt{5})/2]^3 = 2 + \sqrt{5}$ and $[(1 - \sqrt{5})/2]^3 = 2 - \sqrt{5}$. Therefore,
$$\sqrt[3]{2 + \sqrt{5}} + \sqrt[3]{2 - \sqrt{5}} = (1 + \sqrt{5})/2 + (1 - \sqrt{5})/2 = 1.$$

—Leigh Janes.

Q 36. The Buried Treasure

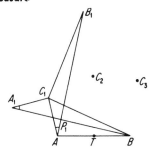

Since triangles AB_1C_1 and A_1BC_1 are congruent, angles C_1AB_1 and C_1A_1B are equal. Then angle AP_1A_1 = angle AC_1A_1 = 90°. Consequently, angle AP_1B = 90°. Thus P_1, and likewise P_2 and P_3, lie on a circle with diameter AB, so the treasure is buried midway between A and B.

—C. F. Pinzka, A.M.M., 65 (June, 1958), 448.

Q 37. Bonus Payments

Let m be the number of men and let x be the fraction of men refusing a bonus. Then the amount paid out is given by $T = 8.15(350 - m) + 10(1 - x)m = 2852.50 + m(1.85 - 10x)$, which will be independent of m only if $x = 0.185$, so that $T = 2852.50$. Both m and $0.185 m$ are integers with $m < 350$, so $m = 200$. It follows that \$1222.50 is paid to the 150 women.

Q 38. Product Simplification

Multiply the product by 1 in the form of $(1/2)\,(3^{2^0} - 1)$ and obtain

$(1/2)\,(3^{2^{n+1}} - 1)$, since $(3^{2^0} - 1)\,(3^{2^0} + 1) = 3^{2^1} - 1$,

$(3^{2^1} - 1)\,(3^{2^1} + 1) = 3^{2^2} - 1$, and so on.

In general, if any base $x > 1$ were used instead of 3, the product would be equal to $[1/(x - 1)](x^{2^{n+1}} - 1)$.

—M.M., 38 (March, 1965), 124.

Q 39. Untangled Strings

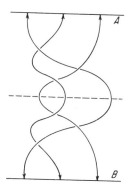

The required pattern of strings may be obtained from the given pattern by taking its reflection in the dotted line.

—N. Grossman, P.M.E.J., 2 (November, 1954), 26.

Q 40. A Difference Equal to a Quotient

Since the quotient is 5, the difference is four times the smaller number. Hence the smaller number is $5/4$ and the larger is $25/4$.

In general, if $x - y = x/y = a$, then $x = a^2/(a - 1)$ and $y = a/(a - 1)$.

Q 41. The Dozing Student

The roots are positive and their arithmetic mean is $-(-20)/20$ while their geometric mean is $(+1)^{1/20}$. Since both means are equal to 1, it follows that all the roots are 1.

—D. S. Greenstein, *A.M.M.*, 63 (September, 1956), 493.

Q 42. Edges of a Polyhedron

Let $n > 3$. A simple polyhedron of $2n$ edges is a pyramid with an n-gon for its base. If the n-gon is folded along a diagonal so that it lies in two planes before its vertices are joined to a point outside the two planes, the resulting polyhedron has $2n + 1$ edges.

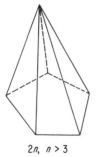 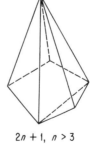 

$2n,\ n > 3$ $\qquad\qquad$ $2n + 1,\ n > 3$ $\qquad\qquad$ 6

Each vertex of a polyhedron is also the vertex of a polyhedral angle with at least three edges, and each edge of the polyhedron serves as an edge of two polyhedral angles. A polyhedron with four vertices is a tetrahedron and has *six* edges. Every other polyhedron has five or more vertices, so has at least $(5)(3)$ edges to its polyhedral angles, and hence at least $(5)(3)/2$ or $7\frac{1}{2}$ edges. Thus seven edges is an impossible case.

—E. P. Starke, A.M.M., 58 (March, 1951), 190.

Q 43. A Simple Congruence

Since $63! - 61! = (63 \cdot 62 - 1)(61!) = 5 \cdot 11 \cdot 71 \ (61!)$, then $63! \equiv 61! \pmod{71}$.

—M.M., 34 (September, 1961), 358.

Q 44. The Triangle is Equilateral

The equation $a^2 + b^2 + c^2 = ab + bc + ca$ is equivalent to $(a - b)^2 + (b - c)^2 + (c - a)^2 = 0$. Hence $a = b = c$, since each term must vanish.

—M. S. Klamkin, M.M., 27 (May, 1954), 287.

Q 45. Simpler Than It Looks

$$\left(\frac{1 \cdot 2 \cdot 4 + 2 \cdot 4 \cdot 8 + 3 \cdot 6 \cdot 12 + \cdots}{1 \cdot 3 \cdot 9 + 2 \cdot 6 \cdot 18 + 3 \cdot 9 \cdot 27 + \cdots} \right)^{1/3}$$

$$= \left(\frac{1 \cdot 2 \cdot 4(1^3 + 2^3 + 3^3 + \cdots)}{1 \cdot 3 \cdot 9(1^3 + 2^3 + 3^3 + \cdots)} \right)^{1/3} = (8\!/\!27)^{1/3} = 2\!/\!3.$$

—Max Beberman, S.S.M., 49 (October, 1949), 588.

Q 46. Covering a Checkerboard

Yes.   All that is necessary is to partition the checkerboard into a closed path one square wide.   The figure on the left shows the symmetrical design by Ralph E. Gomory given in Martin Gardner's *Mathematical Games* column, *Scientific American*, November, 1962, pages 151–152.   The other designs are three of many that will accomplish the same purpose.

The squares lie with alternating colors along the closed paths. The removal of two squares of opposite colors from any two spots along the path will cut the path into two open-ended segments (or one segment if the removed squares are adjacent on the path). Since each segment must consist of an even number of squares, each segment (and therefore the entire board) can be completely covered by dominoes.

Q 47. A Numerical Equality

This is a special case of the algebraic identity

$$a(a + b)(a + 2b)(a + 3b) = (a^2 + 3ab + b^2)^2 - b^4$$

in which we take $a = r^3 + r^2 + r$ [where r is the base (radix) of the scale of notation] and $b = 1$, so that

$$a^2 + 3ab + b^2 = r^6 + 2r^5 + 3r^4 + 5r^3 + 4r^2 + 3r + 1.$$

Since no coefficient exceeds 5, the identity is significant for any base greater than five.

—E. P. Starke, *A.M.M.*, 51 (December, 1944), 590.

Q 48. Book Publishing

The arithmetic mean of the publication years is 13,524/7 or 1932. This middle term is separated from the first term by three common differences, so the first book was published in $1932 - 3(7)$ or 1911.

—*M.M.*, 34 (September, 1961), 372.

Q 49. Three Means

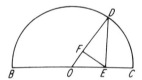

On line segment BC (with $BE = a$ and $EC = b$) as diameter draw a semicircle with center O. At E erect a perpendicular to BC meeting the semicircle at D. Draw OD and to it drop a perpendicular EF from E. Now the radius $OD = (a + b)/2 = A$, and $ED = (ab)^{1/2} = G$. From the similar triangles OED and EFD, $DF/ED = ED/OD$, so $DF = (ED)^2/OD = 2ab/(a + b) = H$. Hence, $G^2 = HA$. Furthermore, $A \geqq G \geqq H$.

—Adrien L. Hess, *S.S.M.*, 61 (January, 1961), 45.

Q 50. Beauty Contest

In the ordering $D–A–E–C–B$, the two in proper position must be adjacent, for otherwise we would have a correct follower after a proper position, implying three in proper positions. This means that $D–A$, $A–E$, $E–C$, or $C–B$ are properly located. The pairs $A–E$ and $E–C$ must be eliminated since a second correct follower could not occur. If $D–A$ is properly placed, the order would have to be either $D–A–B–E–C$ or $D–A–C–B–E$, both of which are ruled out by the comment on the first guess. Hence $C–B$ is in correct position, and $A–E–D–C–B$ is ruled out by the first guess comment, leaving only $E–D–A–C–B$ meeting all requirements.

—J. F. Leetch, *A.M.M.*, 68 (August, 1961), 669.

Q 51. An Equation Involving Series

The principle of composition states that if $a/b = c/d$, then $(a + b)/b = (c + d)/d$. Applying this principle to the given equation,

$$\frac{1^3 + 2^3 + 3^3 + \cdots + (2n)^3}{2^3(1^3 + 2^2 + 3^3 + \cdots + n^3)} = \frac{441}{242}$$

Then, summing the cubes,

$$\frac{(2n)^2(2n + 1)^2/4}{8n^2(n + 1)^2/4} = \frac{441}{242} = \frac{(21)^2}{2(11)^2}$$

$$\frac{(2n + 1)^2}{(n + 1)^2} = \frac{(21)^2}{(11)^2}$$

Clearly, $n = 10$. (The negative square root gives a negative value for n.)

Q 52. Piled Dominoes

If a pile of $k - 1$ rectangular parallelepipeds, $2x$-by-x-by-$x/4$, is placed on another congruent horizontal parallelepiped and slid along until its center of gravity lies above the leading edge of the kth parallelepiped, the center of gravity of the *entire* configuration lies a horizontal distance x/k from the leading edge of the kth parallelepiped.

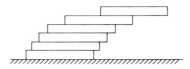

It follows that a columnar pile of n dominoes, with $x = 1$ inch, may be slid between two parallel vertical planes 1 inch apart so that the successive overhangs from the top down constitute the series, $1 + \frac{1}{2} + \frac{1}{3} + \cdots + 1/(n - 1)$, thus forming a half-arch in equilibrium. Since the harmonic series is divergent, the sum of this series will increase with n and the overhang of the top domino over the bottom one may be made any desired amount, depending upon the number of dominoes available.

In order that the *entire* top domino may overhang the bottom one, n must be at least 5, in which case the overhang of the leading edge of the top domino is $1 + \frac{1}{2} + \frac{1}{3} + \frac{1}{4}$ or approximately 2.083 inches.

If the pile be slid so that the diagonals of the dominoes lie in the same vertical plane and the centers of gravity of the various sub-piles from the top down lie over a corner of the leading edge of the supporting domino, then the total overhang will be $[1^2 + (\frac{1}{2})^2]^{1/2}$ or $\sqrt{5}/2$ times that of the first mentioned arrangement.

—*P.M.E.J.*, 1 (April, 1954), 411.

Q 53. Tossing a Die

Consider the throw before the last.   The total after that throw must be 12, 11, 10, 9, or 8.   If the total is 12, the final result must be 13, 14, 15, 16, or 17, with an equal chance for each.   Similarly, if the total is 11, the final result must be 13, 14, 15, 16 with an equal chance for each, and so on.   It is now clear that the most likely final result is 13.

—N. J. Fine, A.M.M., 55 (February, 1948), 98.

If 12 is replaced by $N > 3$, the most likely total is $N + 1$.

Q 54. System of Linear Equations

If $n \geq 3$, the system is dependent and the solution is not unique. Hence $n < 3$.   But the term "system" implies $n > 1$.   Hence $n = 2$. If the equations are

$$ax + (a + d)y = a + 2d$$

$$(a + 3d)x + (a + 4d)y = a + 5d$$

then $x + y = 1$ and $x = -1, y = 2$.

—David Rothman, A.M.M., 70 (January, 1963), 93.

Indeed, any two members of the family

$$(a + 3kd)x + [a + (3k + 1)d]y = a + (3k + 2)d$$

have the same unique solution.

Q 55. Division by Angle Bisector

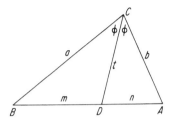

The areas of triangles with equal altitudes are proportional to their bases. Triangles BCD and ACD with bases m and n, respectively, have the same altitude. All points on the bisector of an angle are equidistant from the sides of the angle, so altitudes from D to BC and to AC are equal. Consequently,

$$m/n = \Delta BCD/\Delta ACD = a/b.$$

—W. C. McDaniel, *M.M.*, 44 (November, 1971), 296.

Q 56. Divisibility Probability

The number, 76, formed by the last two digits is divisible by 4. The difference between 73, the sum of the even-placed digits, and $17 + 45$, the sum of the odd-placed digits, is divisible by 11, regardless of the order in which the blanks are filled.   The sum of all the digits, $90 + 45$, is divisible by 9.   It follows that the number is divisible by $(4)\,(11)\,(9)$ or 396.   The probability is 1.

—Prasert Na Nagara, *A.M.M.*, 58 (December, 1951), 700.

Q 57. Equation with No Integer Solutions

Any integer x takes one of the forms $3n$, $3n \pm 1$. If these are substituted in $x^2 - 3y^2 = 17$, the results may be written as

$$3(3n^2 - y^2) = 17 \qquad \text{and} \qquad 3(3n^2 \pm 2n - y^2) = 16,$$

respectively. Since 3 does not divide either 17 or 16, these equations are impossible in integers.

—E. P. Starke, *A.M.M.*, 52 (December, 1945), 580.

Q 58. Son of a Mathematics Professor

Since $f(0) = P$,

$$f(x) = x \cdot q(x) + P \qquad \text{and} \qquad f(A) = A \cdot q(A) + P = A.$$

Consequently, A divides P. Since $P > A$ and P is prime, $A = 1$ year. $x^3 - 3x^2 + 3$ is one of an infinite class of polynomials that the professor could have written.

Q 59. Locus in Space

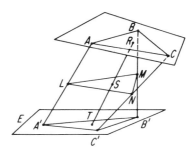

Consider unit particles at the points A, B, C, A', B', C'. Let R be the center of mass of the particles at A, B, C and T the center of mass of the particles at A', B', C'. Now S may be considered as the center of mass of three particles each of mass 2 at L, M, and N; or as the center of mass of two particles each of mass 3 at R and T. So S is the midpoint of RT, and since R is fixed and T varies, the locus of S is a plane parallel to E.

—Daniel Pedoe, *A.M.M.*, 71 (June, 1964), 670.

Q 60. Weather Analysis

There must have been $\frac{1}{2}$ $(6 + 7 - 9)$ or 2 completely clear days, so there were $9 + 2$ or 11 days in the period.

—*M.M.*, 34 (March, 1961), 244.

Q 61. The Steiner-Lehmus Theorem

Using a familiar formula for the length of an angle bisector in terms of the sides,

$$\frac{bc(a + b + c)(b + c - a)}{(b + c)^2} = t_a{}^2 = t_b{}^2 = \frac{ac(a + b + c)(c + a - b)}{(a + c)^2}.$$

This simplifies to

$$c(a + b + c)(a - b)[(a + b)(c^2 + ab) + 3abc + c^3] = 0.$$

Since all factors are positive except $(a - b)$, it follows that $a = b$.

This method of solution is attributed to Jacob Steiner about 1844.

Q 62. The Hula Hoop

Since motion is relative, consider the hoop as fixed and the poor girl whirling around. The original point of contact on the girl traverses the diameter of the hoop twice, and this is the required distance.

—Leo Moser, *A.M.M.*, 66 (December, 1959), 918.

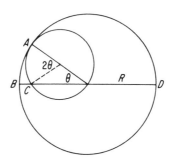

As the girl whirls, the original point of contact C on her waist traverses the diameter BD, since

$$\text{arc } AB = R\theta = (R/2)(2\theta) = \text{arc } AC.$$

Q 63. Octasection of a Circle

Apply the familiar rotation formulas,

$$x = X \cos 45° - Y \sin 45° = (X - Y)/\sqrt{2},$$

$$y = X \sin 45° + Y \sin 45° = (X + Y)/\sqrt{2},$$

to the given equation, and secure,

$$X^4 + kX^3Y - 6X^2Y^2 - kXY^3 + Y^4 = 0.$$

Hence, the graph of the equation is invariant under a rotation through 45°, so the circle is cut into 360/45 or 8 equal parts.

—Norman Anning, *M.M.*, 32 (May, 1959), 285.

The left-hand member of the equation factors into

$$(x + ay)[(a + 1)x + (a - 1)y](ax - y)$$
$$\times [(a - 1)x - (a + 1)y] = 0,$$

where $k = (a^4 - 6a^2 + 1)/a(a^2 - 1)$.

Q 64. Arithmetic Progression Devoid of Powers

The progression 2, 6, 10, $\cdots$, $(4k + 2)$, $\cdots$ can contain no perfect powers whatsoever. For a power of an odd integer is odd, and a power of an even integer must be divisible by 4.

—Azriel Rosenfeld, *A.M.M.*, 62 (March, 1955), 185.

There is also an obvious trivial solution in which the first term is a nonpower and the common difference is zero.

Q 65. Area of Polygon

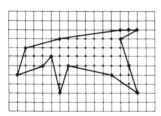

Pick's theorem states that the area of any simple polygon whose vertices are lattice points is given by the formula

$$b/2 + c - 1$$

where b is the number of lattice points on the boundary while c is the number of lattice points inside. Thus the area of the polygon in the figure is $14\frac{1}{2} + 42 - 1$, or 48.

Q 66. A Factorial Equation

We have $n!(n - 1)! = n[(n - 1)!]^2 = m!$ It is evident that $1!\,0! = 1!$ and $2!\,1! = 2!$ are solutions. Otherwise $n < m$, so a solution cannot exist if $m!$ contains a nonsquare factor $> m$. Now for $m > 10$ there are always at least two primes, p and q, which are $> m/2$ and $\leq m$. Then

$$pq \geq \left(\frac{m}{2} + \frac{1}{2}\right)\left(\frac{m + 1}{2} + 2\right) = \frac{m^2}{4} + \frac{3m}{2} + \frac{5}{4} > m.$$

Hence no solution exists for $m > 10$. For $m \leq 10$, the only solution other than the two above is $7!\,6! = 10!$

Q 67. Two Related Triangles

Since $|\sqrt{b} - \sqrt{c}|(\sqrt{b} + \sqrt{c}) = |b - c| < (\sqrt{a})^2 < (b + c) < (\sqrt{b} + \sqrt{c})^2$ it follows that $\sqrt{b} - \sqrt{c} < \sqrt{a} < (\sqrt{b} + \sqrt{c})$.

—Chih-yi Wang, *A.M.M.*, 67 (January, 1960), 82.

Q 68. Maximum Angle in a Circle

The greatest angle is subtended at the point of contact of the smaller of two circles through A and B which touch the given circle

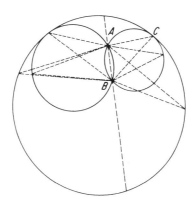

internally. The other circular points on the same side of AB give smaller angles as they are outside the touching circle, while the points on the other side give smaller angles because they are outside the larger touching circle.

—Alan Sutcliffe, M.M., 38 (March, 1965), 124.

Q 69. Determinant of a Magic Square

Let $\begin{matrix} a & b & c \\ d & e & f \\ g & h & i \end{matrix}$ have the magic sum $N = S/3$. Then

$$N = (a + e + i) + (d + e + f) + (g + e + c)$$
$$- (a + d + g) - (c + f + i)$$

$= 3e$, and $S = 9e$. Hence, adding rows and columns,

$$D = \begin{vmatrix} a & b & c \\ d & e & f \\ g & h & i \end{vmatrix} = \begin{vmatrix} a & b & c \\ d & e & f \\ 3e & 3e & 3e \end{vmatrix} = \begin{vmatrix} a & b & 3e \\ d & e & 3e \\ 3e & 3e & 9e \end{vmatrix} = \begin{vmatrix} a & b & e \\ d & e & e \\ 1 & 1 & 1 \end{vmatrix} S.$$

—R. J. Walker, A.M.M., 56 (January, 1949), 33.

Q 70. Five-digit Nonsquares

There are but two sets of five digits, 0 2 4 6 8 and 1 3 5 7 9, which are distinct and congruent modulo 2. The sum of the digits of every perfect square must be congruent modulo 9 to 0, 1, 4, or 7. However, the sum of the digits of the first set is congruent to 2 modulo 9. If the last digit of a perfect square is odd, the penultimate digit must be even. The second set contains no even digit. Hence, no permutation of either set can be a square number.

—A.M.M., 44 (April, 1937), 248.

Q 71. Dissection of Spherical Surface

Inscribe a regular dodecahedron or a regular icosahedron in the sphere and drop perpendiculars from the center to each face. Form isosceles triangles, 60 in number, with the feet of these perpendiculars as vertices and the sides of the respective faces as bases, and centrally project these 60 triangles onto the surface of the sphere as isosceles spherical triangles.

—W. R. Ransom, A.M.M., 40 (February, 1933), 114.

Q 72. Trisector of Side of Triangle

Through B draw a parallel to the median AD meeting CA extended in E. Let M be the midpoint of the median and let CM extended meet BE in F.

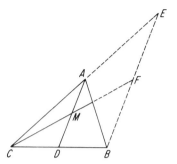

At once, A is the midpoint of CE and F is the midpoint of BE. Then AB and CF are medians of triangle CBE, so they divide each other in the ratio $1:2$.

—Aaron Buchman, S.S.M., 50 (December, 1950), 757.

Q 73. A Factorization

Factoring variously,

$$
\begin{aligned}
a^{15} + 1 &= (a^3 + 1)(a^{12} - a^9 + a^6 - a^3 + 1) \\
&= (a + 1)(a^2 - a + 1)(a^{12} - a^9 + a^6 - a^3 + 1) \\
&= (a^5 + 1)(a^{10} - a^5 + 1) \\
&= (a + 1)(a^4 - a^3 + a^2 - a + 1)(a^{10} - a^5 + 1).
\end{aligned}
$$

Inspection reveals that $a^2 - a + 1$ does not divide $a^4 - a^3 + a^2 - a + 1$ hence it must divide $a^{10} - a^5 + 1$. So we write, $a^{10} - a^5 + 1 = (a^{10} - a^9 + a^8) + (a^9 - a^8 + a^7) - (a^7 - a^6 + a^5) - (a^6 - a^5 + a^4) - (a^5 - a^4 + a^3) + (a^3 - a^2 + a) + (a^2 - a + 1)$. Therefore, $a^{15} + 1 = (a + 1)(a^4 - a^3 + a^2 - a + 1)(a^2 - a + 1)(a^8 + a^7 - a^5 - a^4 - a^3 + a + 1)$.

—M.M., 27 (May, 1954), 287.

Q 74. A Curious Number

If $(N/5)(N/7) = N$, then $N(N - 35) = 0$, so $N = 35$.

Otherwise.   If, instead of multiplying $\frac{1}{5}$ of the number by $\frac{1}{7}$ of the number, the number were multiplied by itself, the product would be 35 times as great as by the other method, so the number is 35.

In general, if a number is equal to $\Pi\, 1/a_i$ of itself, the number is $\Pi\, a_i$.

Q 75. Two Regular Hexagons

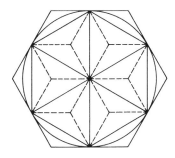

Consider an inscribed regular hexagon with vertices at the mid-points of the circumscribed hexagon.   By drawing the radii of the inscribed hexagon and then joining the centroids of the six equilateral triangles thus formed to the vertices of their respective triangles, the configuration is divided into 24 congruent triangles.   Eighteen of these lie in the inscribed hexagon, so the areas of the inscribed and circumscribed hexagons are in the ratio of 18:24 or 3:4.

—M.M., 35 (March, 1962), 70.

Q 76. General Terms of Series

With either series, any number whatsoever could be used as a sixth term.   Consequently, any number of general formulas could be found for each series.   Two that are evident upon inspection are:

(a)   0, 3, 26, 255, 3124, $\cdots$, $(n^n - 1)$;

(b)   1, 2, 12, 288, 34560, $\cdots$, $[(1!)(2!)(3!)\cdots(n!)]$.

Q 77. Heat Flow

If four such sheets are superimposed so that a 100° edge appears on each of the four sides, the average temperature of each side is 25°. Hence, the temperature of the middle of each sheet is 25°.

—Leo Moser, *M.M.*, 24 (May, 1951), 273.

This procedure may be generalized to prove: If the edges of a regular n-gonal metal sheet are kept respectively at temperatures t_i degrees, $i = 1, 2, 3, \cdots, n$, then the temperature of the center of the sheet is $(1/n)\sum_{i=1}^{n} t_i$ degrees.

Q 78. Does This Make Sense?

If $\frac{1}{4}$ of 20 is 6, the computation must be in the duodecimal scale, which has the base twelve.   Hence $\frac{1}{5}$ of 10 is $2\frac{2}{5}$, since $10_{12} = 12_{10}$.

—*M.M.*, 31 (January, 1958), 178.

Q 79. Envelope into Tetrahedrons

Since the areas of congruent tetrahedrons are equal, the envelope must be cut so that the two pieces will have equal areas.   Hence, the cut must pass through the center of the rectangle.   First strongly crease the envelope along its diagonals and along the line through the center which is perpendicular to the longer side.   Then the cut may be made in one of four ways: along a diagonal, along the perpendicular to the long side, at an angle through the center cutting the long sides, or through the center cutting the short sides.   In each case, when

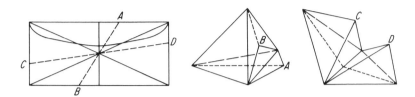

the cut halves are folded out along the diagonals and in along the edges and the long side perpendicular so as to join the extremities of the cut, two congruent isosceles tetrahedrons are formed.

The single exception occurs when the envelope is square. Then instead of tetrahedrons, two square envelopes result from the cutting and folding.

—*A.M.M.*, 56 (June, 1949), 410.

Q 80. A Hammy Cryptarithm

Let $FRY = x$ and $HAM = y$, then

$$7(1000x + y) = 6(1000y + x)$$
$$6994x = 5993y$$
$$538x = 461y$$

Since the numerical coefficients are relatively prime, it follows that $x = FRY = 461$ and $y = HAM = 538$.

Q 81. Sheep Buyers

If x is the number of cattle, y the number of sheep, z the price of the lamb, we have $x^2 = 10y + z$ with y odd and $z < 10$. But the penultimate digit of a square is odd if and only if the last digit is 6. Thus $z = 6$, and the luckier son should hand over $2.

—Irving Kaplansky, *A.M.M.*, 51 (March, 1944), 166.

Q 82. A Constant Volume

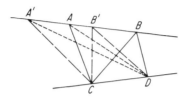

Let the segments move successively, so that one segment is stationary while the other, say AB, moves to a new position $A'B'$. The area of the variable triangle ABC is constant since its altitude does not change. Moreover, the distance from D to the plane of ABC is unchanged. Since the determining factors (base and altitude) remain constant, the volume of the tetrahedron is unaltered.

—Leon Bankoff, *P.M.E.J.*, 1 (November, 1952), 281.

For other short proofs, see Nathan Altshiller-Court, *Modern Pure Solid Geometry*, Macmillan Co. (1935), page 87, where the proposition is called Steiner's Theorem.

Q 83. A Repeating Decimal

The first period of $\frac{1}{7}$ is 0.142857. Upon dividing this period repeated seven times by 7 until the division is exact, the first period of $\frac{1}{7^2}$ is obtained quickly. Thus

7)0.142857 142857 142857 142857 142857 142857 142857
0.020408 163265 306122 448979 591836 734693 877551

—Dewey C. Duncan, *M.M.*, 25 (March, 1952), 224.

Q 84. A Radical Equation

If $a + b + c = 0$, then $a^3 + b^3 + c^3 = 3abc$, so

$$(6x + 28) - (6x - 28) - 8 = 3[(6x + 28)(6x - 28)(8)]^{1/3}$$

$$48 = 6(36x^2 - 784)^{1/3}$$

$$512 = 36x^2 - 784$$

$$x^2 = 36$$

$$x = \pm 6$$

—Frederic E. Nemmers, *S.S.M.*, 41 (March, 1941), 291.

Q 85. The Prize Contest

One may count paths "backwards" from the N.   In counting the left half of the array, including the center column, there are two choices for each backward step.   Thus this portion yields 2^{12} paths. Doubling this number and subtracting the center column to keep from counting it twice, yields $2^{13} - 1$ or 8191 paths.

—J. F. Leetch, *A.M.M.*, 68 (March, 1961), 296.

Q 86. A Constant Sum

Let the vertices and other point be $A(1, 0, 0)$, $B(0, 1, 0)$, $C(0, 0, 1)$, $P(x, y, z)$.   The inscribed circle is the intersection of a sphere $x^2 + y^2 + z^2 = c_1$ and a plane $x + y + z = c_2$.

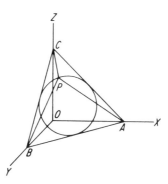

Hence, $(PA)^2 + (PB)^2 + (PC)^2 =$

$(x - 1)^2 + y^2 + z^2 + x^2 + (y - 1)^2 + z^2 + x^2 + y^2 + (z - 1)^2$

$= 3c_1 - 2c_2 + 3 = $ a constant.

—Leo Moser, *A.M.M.*, 56 (March, 1949), 180.

This proof holds for any circle concentric with the incircle.

Q 87. Two Incompatible Integers

Expressed dyadically (that is, to the base two),

$$2^x + 1 = 100\cdots001 \text{ and } 2^y - 1 = 111\cdots111.$$

Now attempt the division

$$111\cdots111 \ ) \ 100\cdots\cdots\cdots\cdots001 \ (\ 1$$
$$\underline{111\cdots111}$$
$$100\cdots001$$

The division will be exact if and only if the final remainder consists of as many 1's as the divisor, which by hypothesis has more than

two 1's.   But the successive remainders consist only of two 1's and
possible intervening zeros, so the division is never exact.

—Dewey C. Duncan, *S.S.M.*, 36 (March, 1936), 321.

Q 88. A Multiple-choice Question

Since the difference of the two terms on the left side of the equation
is the odd number 41, one of the terms must be odd and the other
even.   Since $104y$ is even, $187x$ must be odd, so x is odd.   Therefore
$x = 314$, $y = 565$ cannot satisfy the equation.

—Dale Woods, *S.S.M.*, 64 (March, 1964), 242.

Q 89. Concurrency of Medians

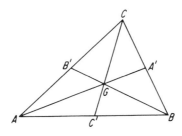

The converse of Ceva's theorem states: If three points taken on
the sides of a triangle divide these sides into six such segments that
the product of three segments having no common end is equal to
the product of the remaining three, then the lines joining the three
points to the opposite vertices of the triangle are concurrent.

Since the medians bisect the sides, $(AB')(CA')(BC') = (B'C)(A'B)(C'A)$, so the medians are concurrent.

—M.M., 24 (November, 1950), 114.

Furthermore, by a corollary of Ceva's theorem,

$$CG/GC' = CB'/B'A + CA'/A'B = 1 + 1$$

so $CG = 2GC'$ and $CG = (2/3)CC'$.   That is, the medians intersect at a point which is $2/3$ the length of each median from its vertex.

Q 90. A Surprising Square

If the base of the system of numeration is B, $B > 1$, then 11111 may be written $B^4 + B^3 + B^2 + B + 1$.   Now

$$(B^2 + B/2)^2 < B^4 + B^3 + B^2 + B + 1 < (B^2 + B/2 + 1)^2.$$

If

$$(B^2 + B/2 + 1/2)^2 = B^4 + B^3 + B^2 + B + 1$$

then

$$B^2/4 - B/2 - \tfrac{3}{4} = 0$$

$$(B - 3)(B + 1) = 0$$

and for $B = 3$, we have $11111 = (102)^2$.

—V. Thebault, N.M.M., 15 (December, 1940), 149.

Q 91. Polygon Inscribed in Ellipse?

If such a polygon existed, then its circumcircle would cut the ellipse in more than four points—the vertices of the polygon.   This is impossible.

—M. S. Klamkin, M.M., 34 (September, 1960), 58.

Q 92. Telephone Call to Sinkiang

Since

$$S = 0.1 + 0.02 + 0.003 + 0.0004 + \cdots$$

$$0.1S = \qquad 0.01 + 0.002 + 0.0003 + \cdots$$

$$0.9S = 0.1 + 0.01 + 0.001 + 0.0001 + \cdots$$

$$= 0.1/(1 - 0.1) = 1/9$$

So $S = {}^{10}\!/_{81} = 0.123456790123456790\cdots$, a periodic decimal not containing the digit 8.

—Norman Anning, *M.M.*, 29 (January, 1956), 173.

Q 93. The Stock Pen

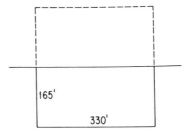

If he built a $2\frac{1}{2}$ acre pen on the open plain, a square pen would enclose the area with a minimum perimeter. That is, a pen $[(\frac{5}{2})(43{,}560)]^{1/2}$ or 330 feet square. The cliff may be thought of as bisecting such a square with a no cost wall. Hence, minimum cost would be achieved with a fencing 330 feet long and 165 feet on each end.

Q 94. Five Simultaneous Linear Equations

Add the five equations and divide through by 4 to obtain

$$x + y + z + u + v = 3.$$

Subtract each of the original equations in order from this last equation to get $v = -2$, $x = 2$, $y = 1$, $z = 3$, $u = -1$.

Q 95. An Almost Universal Theorem

The theorem is $(n - 5)(n - 17)(n - 257) \neq 0$.

—Leo Moser, *M.M.*, 25 (September, 1951), 49.

Q 96. Angle Trisectors

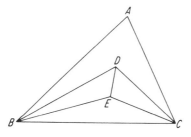

Since E is the incenter of triangle BCD, then DE bisects angle BDC.

—C. F. Pinzka, *M.M.*, 34 (January, 1961), 182.

Q 97. Units' Digits of Fibonacci Series

Yes.   The series of units' digits, $1\,1\,2\,3\,5\,8\,3\,1\cdots$, is formed by suffixing the units' digit of the sum of a pair of digits and continuously repeating the process.

Two odd terms and one even term alternate in the series.   There are only $5(5)$ ordered pairs of odd digits, so after not more than $3(25)$ or 75 operations, one of these pairs must reappear to start a new cycle. Since the sum (or difference) of two digits is unique, when a pair reappears it must be the starting pair rather than one in the interior of the sequence.   Indeed, after 60 additions a new cycle of 60 digits is started, thus $1\,1\,2\,3\,5\,8\,3\,1\,4\,5\,9\,4\,3\,7\,0\,7\,7\,4\,1\,5\,6\,1\,7\,8\,5\,3\,8\,1\,9$ $0\,9\,9\,8\,7\,5\,2\,7\,9\,6\,5\,1\,6\,7\,3\,0\,3\,3\,6\,9\,5\,4\,9\,3\,2\,5\,7\,2\,9\,1\,0^*1\,1\cdots$. The same basic argument holds in any system of numeration, and a similar one holds for the units' digits of the series $A_{n+k} = A_n + A_{n+1} + \cdots + A_{n+k-1}$.

Q 98. Related Cubics

Since the coefficient of x^2 is 0, it follows that $a + b + c = 0$.   Hence $b + c = -a$, $c + a = -b$, and $a + b = -c$, so we seek an equation whose roots are $-1/a$, $-1/b$, and $-1/c$, the negative reciprocals of the given roots.

We need merely write the coefficients in reverse order, and change the signs of the coefficients in even position, thus obtaining

$$rx^3 - qx^2 - 1 = 0.$$

—Alan Wayne, *S.S.M.*, 48 (June, 1948), 492.

Q 99. A Skeleton Product

No one of the units' digits of the factors can be zero. $(4)(6)(8) = 192$ and $(2)(4)(6) = 48$. $(87)^{1/3} = 4.4+$ and $(88)^{1/3} = 4.4+$. Therefore, the product is $(442)(444)(446) = 87,526,608$.

—*M.M.*, 37 (November, 1964), 360.

Q 100. Inscribed Decagons

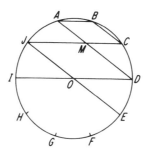

Diameters ID and JE are respectively parallel to the sides AB and BC of the regular decagon and to sides JC and AD of the star decagon. Hence $ABCM$ and $JMDO$ are rhombuses, so $AD - BC = AD - AM = MD = JO$, a radius of the circle.

Q 101. The Handshakers

Before any handshakes have occurred, the number of persons who have shaken hands an odd number of times is zero.   The first handshake will produce two "odd persons."   From then on handshakes will occur between either two even persons, two odd persons, or one odd and one even person.   Each even-even shake increases the number of odd persons by two.   Each odd-odd shake decreases the number of odd persons by two.   Each odd-even shake changes an odd person to even and an even person to odd, leaving the number of odd persons unchanged.   Therefore, there is no way that the even number of odd persons can shift its parity; it must always be an even number.

—Gerald K. Schoenfeld, in Martin Gardner's *2nd Scientific American Book of Mathematical Puzzles and Diversions*, Simon and Schuster, Inc., New York, 1961, page 60.

Q 102. A Fast Deal

We need be concerned only with the order of the cards drawn. The permutations of any five numbers are 5!, so the probability is $\frac{1}{120}$.

—J. R. Ziegler, *M.M.*, 23 (May, 1950), 278.

Q 103. A Perpendicular Bisector

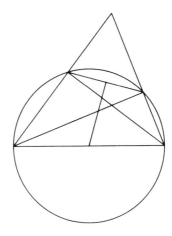

A side of a triangle is the diameter of a circle passing through the feet of the altitudes to the other two sides.   The join of these feet is a chord of the circle, so its perpendicular bisector passes through the center of the circle, that is, the midpoint of the third side.

—Aaron Buchman, *S.S.M.*, 47 (May, 1947), 490.

Q 104. Condition for Divisibility

Let $f(x) = x^2 - x + a$ and $g(x) = x^{13} + x + 90$.   Then

$$f(0) = a. \qquad f(1) = a, \qquad g(0) = 90, \qquad g(1) = 92,$$

so a must divide 2, the highest common factor of 90 and 92.   Also,

$f(-1) = a + 2$ and $g(-1) = 88,$ \qquad so a is not 1 or -2.

$f(-2) = a + 6$ and $g(-2) = -8104,$ \qquad so a is not -1.

Hence $a = 2$ and

$(x^{13} + x + 90)/(x^2 - x + 2)$

$$= x^{11} + x^{10} - x^9 - 3x^8 - x^7 + 5x^6$$
$$+ 7x^5 - 3x^4 - 17x^3 - 11x^2 + 23x + 45.$$

—L. E. Bush, *A.M.M.*, 71 (June, 1964), 640.

Q 105. The Farmer's Dilemma

The average cost per animal is \$1.   Each calf's cost differs from the average by $+\$9$, each lamb's cost by $+\$2$, and each pig's cost by $-\$\frac{1}{2}$.   Hence for each calf he must have 18 pigs and for each lamb he must have 4 pigs.   Therefore, since $5(1 + 18) + (1 + 4) = 100$, he must buy 5 calves, 1 lamb and 94 pigs.

—B. E. Mitchell, *M.M.*, 26 (January, 1953), 153.

Q 106. A Factored Integer

$$1,000,027 = (100)^3 + (3)^3 = (100 + 3)(10,000 - 300 + 9)$$
$$= (103)(9709) = (103)(7)(1387) = (103)(7)(1460 - 73)$$
$$= (103)(7)(73)(19).$$

Q 107. A Folded Card

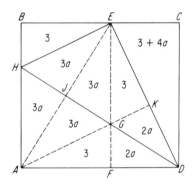

If $\triangle BEH = 3$, let $\triangle CED = 3 + 4a$, then $\triangle EHD = 3 + 8a$. Draw EF perpendicular to AD cutting HD in G. Through G draw AK and draw AE cutting HD in J. From considerations of symmetry in $HEDA$, the rectangle is now divided into ten right triangles, and $\triangle EGK = \triangle AGF = \triangle BEH = 3$. $\triangle CED = \triangle EFD$, so $\triangle GKD = \triangle GFD = 2a$. Hence each of the four triangles in the rhombus $AHEG = 3a$.

From two pairs of similar triangles,

$$\triangle CED/\triangle BEH = (DE)^2/(EH)^2 = \triangle JED/\triangle JEH,$$

so

$$(3 + 4a)/3 = (3 + 5a)/3a$$

$$4a^2 - 2a - 3 = 0$$

$$4a = 1 + \sqrt{13}$$

Consequently, the area of the largest triangle, $\triangle EHD$, is $8a + 3$ or $5 + 2\sqrt{13} \doteq 12.2111$ square inches.

After some more involved computations, we also find that $BE \doteq 3.528$ inches, $EC \doteq 2.708$ inches, and $CD \doteq 5.617$ inches.

—R. R. Rowe, *Civil Engineering-A SCE*, 18 (February, 1948), 70.

Q 108. A Unique Square

If $n(n + 2)(n + 4)(n + 6) = m^2$, then $(n^2 + 6n + 4)^2 = m^2 + 16$. But only 0 and 9 are squares of the form $a^2 - 16$, and since m^2 is odd, the square sought must be $9 = (-3)(-1)(1)(3)$.

—David L. Silverman, *M.M.*, 38 (January, 1965), 60.

Q 109. A Neat Inequality

The arithmetic mean of a set of numbers, not all equal, is greater than their geometric mean, so

$$n = \frac{n^2}{n} = \frac{1 + 3 + 5 + 7 + \cdots + (2n - 1)}{n}$$

$$> [1 \cdot 3 \cdot 5 \cdot 7 \cdots (2n - 1)]^{1/n}$$

Whereupon, $n^n > 1 \cdot 3 \cdot 5 \cdot 7 \cdots (2n - 1)$.

—Frederic E. Nemmers, *S.S.M.*, 40 (June, 1940), 586.

Q 110. A Cosine Sum

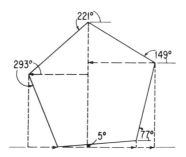

Considered as directed lines, the projections of the sides of any polygon on a line in its plane sum to zero. Since the exterior angle of a regular pentagon is 72°, the terms in the given sum are the orthogonal projections of the unit sides of a regular pentagon on a line with which one of the sides makes an angle of 5°. Consequently,

$$\cos 5° + \cos 77° + \cos 149° + \cos 221° + \cos 293° = 0.$$

—M. S. Klamkin, *M.M.*, 28 (May, 1955), 293.

Q 111. Quantity Divisible by 9

We may write

$$f(x) = \frac{x^{12} - 1}{x^2 - 1},$$

so

$$f(2i) = \frac{(2i)^{12} - 1}{(2i)^2 - 1} = \frac{4096 - 1}{-4 - 1} = -819 = -9(91).$$

—*M.M.*, 26 (May, 1953), 287.

Q 112. Triangular Numbers in Scale of Nine

Triangular numbers have the form $n(n + 1)/2$, so 1 is a triangular number in any scale of notation. Each member of the series is derived from the previous member by multiplying by the base and adding 1. Thus operating in the decimal system, for the scale of nine,

$$9n(n + 1)/2 + 1 = (3n + 1)(3n + 2)/2,$$

a triangular number.

—Helen A. Merrill, *A.M.M.*, 39 (March, 1932), 179.

In general, in a system with base $(2k + 1)^2$, annex $k(k + 1)/2$ to $n(n + 1)/2$ and secure $[(2k + 1)n + k][(2k + 1)n + k + 1]/2$.

Q 113. Relative Polyhedral Volumes

A plane through the midpoints of the three edges of a tetrahedron issuing from one vertex will cut off a smaller tetrahedron with a volume $1/2^3$ that of the larger similar one (since their edges are in the ratio of 1:2). Four of these smaller tetrahedra together have a volume equal to one-half that of the larger tetrahedron.

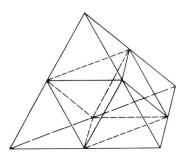

If a regular tetrahedron is truncated in this fashion at all four vertices, the residual solid has four congruent equilateral triangular faces from the original faces and four from the truncations, so it is a regular octahedron with edges equal to those of the smaller tetrahedrons and a volume equal to one-half that of the larger tetrahedron.

It follows that a smaller tetrahedron has a volume one-fourth that of the octahedron.   Also, from the truncations, it is evident that the dihedral angles of the regular tetrahedron and the regular octahedron are supplementary.

Q 114. The Space Fillers

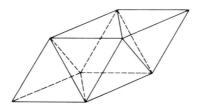

A cube can be continuously deformed until the edges issuing from one vertex make angles of 60° with each other.   The joins of the extremities of these edges together with the edges form a regular tetrahedron.   The edges issuing from the opposite vertex likewise determine a regular tetrahedron.   Since one join is in each face of the parallelepiped (the deformed cube), the aforesaid joins together with six edges of the parallelepiped determine a regular octahedron.   It follows that a cubical network (which is space filling) can be continuously transformed into a tessellation of n regular octahedrons and $2n$ regular tetrahedrons.

Or, as might be inferred from the previous problem, if the edges of an infinite tetrahedron be divided into equal parts and planes parallel to the faces be passed through the sectioning points, the tetrahedron (and hence space) will be divided into regular octahedrons and tetrahedrons.

Q 115. Simple Simplification

The sum of the digits of the numerator is 30, so the numerator is divisible by 3. The difference of the sums of the alternate digits of the denominator is $(32 - 21)$ or 11, so the denominator is divisible by 11. Therefore,

$$\frac{116,690,151}{427,863,887} = \frac{38,896,717\,(3)}{38,896,717\,(11)} = \frac{3}{11}.$$

Q 116. Sine Sum

The triangle of maximum perimeter inscribed in a circle is equilateral, for which the sides $a = b = c = R\sqrt{3}$, where R is the radius of the circle. Hence, for the general triangle, $a + b + c \leqslant 3R\sqrt{3}$. But $a = 2R\sin A$, $b = 2R\sin B$, and $c = 2R\sin C$, so $\sin A + \sin B + \sin C \leqslant 3\sqrt{3}/2$.

—Leon Bankoff

129

Q 117. Two Ferry Boats

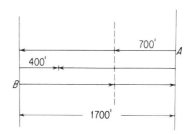

Ferry boat A left one shore, traveled 700 feet and met B. Together they had traveled the width of the river. A continued across the river to the opposite shore and back 400 feet, where it met B again. Together they had then traveled a total of three times the width of the river. As their speeds were constant, A traveled three times 700 feet, or 2100 feet. The width of the river was 400 feet less than the distance A traveled, that is, 1700 feet.

—W. C. Rufus, A.M.M., 47 (February, 1940), 111.

Q 118. Togetherness at Meals

There would be $\frac{6}{2}$ or 3 distinct pairs with which to bracket Albert, so there must be 3 meals in a cycle. Using the initials of the names, let us first seat the family in alphabetical order and then break the circle at Albert's right and straighten out the sequence for consideration. Then take those in an even position and move them to the right, preserving their order, and repeat the operation twice. The third move returns the seating to the original alphabetical order.

Thus,

$$A \ B \ C \ D \ E \ F \ G$$
$$A \ C \ E \ G \ B \ D \ F$$
$$\underline{A \ E \ B \ F \ C \ G \ D}$$
$$A \ B \ C \ D \ E \ F \ G$$

Now let us examine the bracketing pairs for each person, and incidentally demonstrate that the device used does the job.

A—BG	B—AC	C—BD	D—CE	E—DF	F—EG	G—FA
CF	GD	AE	BF	CG	DA	EB
ED	EF	FG	GA	AB	BC	CD

Thus not only does each person sit by every other person exactly once during the three meal cycle, but at no time during the cycle does any person have the same bracketing pair of neighbors as any other person during the cycle.   That is, the bracketing pairs are the 21 that are the $C(7, 2)$.

Q 119. Sum of Digits

Pair the integers a and $(10^n - 1 - a)$, $a \geqq 0$.   Each pair has a digit sum of $9n$, and the number of pairs is $10^n/2$.   Hence, the required sum is $9n(10^n/2)$.

—Leo Moser, *M.M.*, 26 (March, 1953), 225.

Q 120. Feeding Three Truck-drivers

The purchases of the first two truck-drivers establish the equations

$$4s + c + 10d = 169 \tag{1}$$

and

$$3s + c + 7d = 126 \tag{2}$$

Then we have

$$2 \text{ times } (1): \quad 8s + 2c + 20d = 338 \qquad (3)$$

$$3 \text{ times } (2): \quad 9s + 3c + 21d = 378 \qquad (4)$$

$$(4)-(3): \quad s + c + d = 40 \text{ cents}$$

which was paid by the third truck-driver.

—*S.S.M.*, 66 (June, 1966), 561.

Q 121. Overlapping Squares

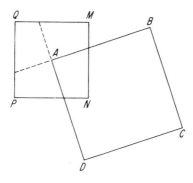

Extend the sides of angle A, thus dividing the square $MNPQ$ into four congruent quadrilaterals. Thus the common area is one-fourth of the area of $MNPQ$. This result is independent of the ratio into which MN is divided and of the size of $ABCD$ provided that $AB \geqq MP/2$.

Q 122. Solution without Expansion

Let $x = y/12$, then the equation becomes

$$(y - 1)(y - 2)(y - 3)(y - 4) = 120 = 2 \cdot 3 \cdot 4 \cdot 5.$$

Any integer root will convert the left-hand member into the product of four consecutive integers. $y = -1$ and $y = 6$ are two such roots.

Now applying the relationships between the coefficients and the roots of the derived equation:

$$(-1)(6)r_1r_2 = -120 + (-1)(-2)(-3)(-4)$$

or

$$r_1r_2 = 16$$

$$-(-1 + 6 + r_1 + r_2) = -1 - 2 - 3 - 4$$

or

$$r_1 + r_2 = 5.$$

That is, r_1 and r_2 are roots of $y^2 - 5y + 16 = 0$, namely, $(5 \pm i\sqrt{39})/2$. It follows that the four roots of the original equation are $-1/12$, $1/2$, and $(5 \pm i\sqrt{39})/24$.

Q 123. Problem in Primes

It is required to find a three-digit and a one-digit number whose product has four digits, with the condition that only 2, 3, 5 or 7 be used. There are only four ways to meet this requirement: $3(775) = 2325$, $5(555) = 2775$, $5(755) = 3775$, and $7(325) = 2275$.

Since no three-digit number appears with more than one multiplier, the multiplier sought must consist of two identical digits. The unique solution is

$$
\begin{array}{r}
775 \\
33 \\
\hline
2325 \\
2325 \\
\hline
25575
\end{array}
$$

—W. E. Buker, *A.M.M.*, 43 (October, 1936), 499.

Q 124. Intersecting Great Circles

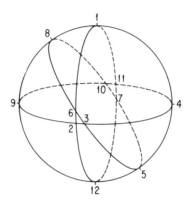

Place any number b on any point and place its "complement," $n(n - 1) + 1 - b$ on the diametrically opposite point. Continue in this way until all the numbers (and all the points) are exhausted. Since each circle contains $n - 1$ pairs of diametrically opposite points, the numbers on each circle will total to $[n(n - 1) + 1](n - 1)$.

—Leo Moser, *M.M.*, 25 (November, 1951), 114.

Q 125. Costly Club

With 50 percent more to share the expense, the cost per person would have been $\frac{2}{3}$ as great. Thus, $100 is one-third and the initial individual cost was $300.

—*M.M.*, 32 (March, 1959), 229.

Q 126. Binomial Coefficients

For $y = 3$, we have $2C(n, k) = C(n, k + 1)$ and $3C(n, k) = C(n, k + 2)$. These equations simplify to $n = 3k + 2$ and $3(k + 1)(k + 2) = (n - k)(n - k - 1)$, respectively. Solving simultaneously and discarding the negative root, $k = 4$, $n = 14$. That is, in the binomial expansion to the fourteenth power, the fifth, sixth, and seventh coefficients are 1001, 2002, 3003, and the eighth coefficient $3432 \neq 4(1001)$. Since this solution is unique for $y = 3$, there can be no solution for $y > 3$.

Q 127. Quantity Divisible by 8640

The product of n consecutive integers is divisible by n. Furthermore, the product of four consecutive integers is divisible by 2^3. Now

$$N = x^9 - 6x^7 + 9x^5 - 4x^3$$
$$= [(x - 2)(x - 1)x][(x - 1)x(x + 1)][x(x + 1)(x + 2)]$$
$$= [(x - 2)(x - 1)x(x + 1)(x + 2)][(x - 1)x][x(x + 1)]$$
$$= [(x - 2)(x - 1)x(x + 1)][(x - 1)x(x + 1)(x + 2)]x.$$

From the first arrangement of the factors, the quantity in each bracket is divisible by 3 so N is divisible by 3^3. From the second arrangement N is divisible by 5. From the third arrangement the quantity in each bracket is divisible by 2^3 so N is divisible by 2^6. Therefore $2^6 \cdot 3^3 \cdot 5$ or 8640 divides N for all integer values of x.

Q 128. Dissected Pentagon

If each side of the square is 2, the area of the pentagon is 5, so that each leg of the resulting isosceles triangle must be $\sqrt{10}$. This is the length of each of the long diagonals of the pentagon, which may be drawn from its apex. Thus if we cut along a long diagonal, the two sides of the cut will form the legs of the final triangle. Then the vertices of the uncut angles of the pentagon should come together inside the final triangle and the pairs of sides issuing from the ends of the diagonal must match. Hence the second cut must be from either

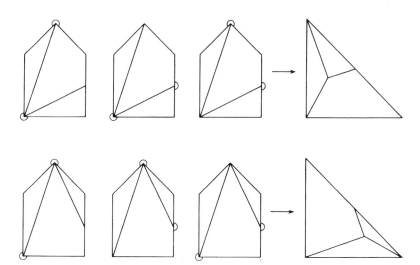

end of the long diagonal of the pentagon and must bisect the remote side. There are thus two distinct solutions, not counting mirror images. In each case, the pieces may be swung from their initial to their final positions as if they were hinged at any two of the three points where cuts end.

—W. Fitch Cheney.

Q 129. Infinite Product

Let $N = 3^{1/3}3^{2/9}3^{3/27}\cdots = 3^{1/3+2/9+3/27+\cdots+n/3^n+\cdots} = 3^M$. Then

$$M/3 = \tfrac{1}{3^2} + \tfrac{2}{3^3} + \tfrac{3}{3^4} + \cdots + (n-1)/3^n + \cdots$$

So

$$(1 - \tfrac{1}{3})M = \tfrac{1}{3} + \tfrac{1}{3^2} + \tfrac{1}{3^3} + \cdots + \tfrac{1}{3^n} + \cdots$$
$$= (\tfrac{1}{3})/(1 - \tfrac{1}{3}) = \tfrac{1}{2}.$$

Hence $N = 3^{3/4}$ or $\sqrt[4]{27}$.

—J. F. Arena, *S.S.M.*, 46 (October, 1946), 678.

Q 130. Never a Square

We have $(n^2 + n)^2 = n^4 + 2n^3 + n^2 < n^4 + 2n^3 + 2n^2 + 2n + 1 < n^4 + 2n^3 + 3n^2 + 2n + 1 = (n^2 + n + 1)^2$. So the given number lies between two consecutive squares.

—Leo Moser, *M.M.*, 37 (January, 1964), 62.

Indeed,

$$n^4 + 2n^3 + 2n^2 + 2n + 1$$
$$= (n^4 + 2n^3 + n^2) + (n^2 + 2n + 1) = (n^2 + 1)(n + 1)^2.$$

Q 131. Chords of a Circle

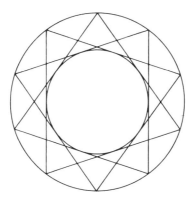

The lines are equal chords tangent to a circle concentric with the given circle. If more than two lines passed through one point, there would be more than two tangents from an external point to a circle, which is impossible.

—Brother U. Alfred, *M.M.*, 35 (May, 1962), 193.

Q 132. Nine-digit Determinants

When two rows of a determinant are interchanged, the sign of the determinant is changed.   When the rows of a 3-by-3 determinant are permuted, three positive and three negative determinants equal in absolute value are obtained.   Hence, the 9! determinants fall into 9!/6 groups each of which sums to zero.

—*M.M.*, 36 (January, 1963), 77.

Q 133. Six Common Points

If the graphs of the quadratic equations have more than $(2)(2)$ or 4 points in common, the equations must be degenerate with a common factor.   The given equations may be written in the forms

$$(x + 2y - 3)(2x - y) = 0$$

$$(x + 2y - 3)(3x + y + 2) = 0.$$

Hence all points on the line $x + 2y - 3 = 0$ lie on the graphs of both quadratics.   For example, the coordinates of the points $(-1, 2)$, $(1, 1)$, $(0, \frac{3}{2})$, $(3, 0)$, $(4, -\frac{1}{2})$, $(5, -1)$ satisfy both equations.

Q 134. A Shuffled Deck

The number of red cards in the top 26 must always equal the number of black cards in the bottom 26.   Hence by the rules of logic, the statement is correct no matter what follows the word "then."

—Leo Moser, *M.M.*, 26 (January, 1953), 167.

Q 135. Equal Angles

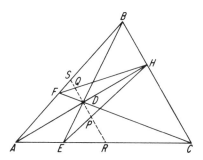

Through D draw a parallel to BC cutting HE, HF, AC, and AB in points P, Q, R, S, respectively. Then

$$\frac{DP}{DR} = \frac{BH}{BC}, \qquad \frac{DS}{DQ} = \frac{BC}{CH}, \qquad \text{and} \qquad \frac{DR}{DS} = \frac{CH}{BH}.$$

Multiplying these three equalities we have $DP = DQ$. Thus in the triangle HPQ the altitude bisects the base PQ, hence HPQ is isosceles and HA bisects angle FHE to make angles AHE and AHF equal.

—Nathan Altshiller Court, *M.M.*, 37 (November, 1964), 338.

Q 136. A Consistent System

The sum of the second and third equations is $x + y + k(x + y) = 5$. Hence from the first equation, $k = 4$.

Q 137. Diophantine Equation

Put $x + y = a^2$ and $x - y = a$, then $x = a(a + 1)/2$ and $y = a(a - 1)/2$. Since each numerator is the product of an odd and an even integer, x and y are integers when a is an integer.

—L. E. Bush, *A.M.M.*, 61 (October, 1954), 548.

x and y are consecutive triangular numbers.

Q 138. Dissection of Triangle into Two Similar Triangles

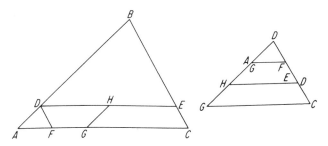

Take D, E, F on AB, BC, CA, the sides of any triangle ABC, such that $AD/AB = CE/CB = AF/FC = \frac{1}{5}$. Let G be on AC with $AG = 2\,AF$, and let H be the midpoint of DE. Then the segments DE, DF, GH represent three cuts. One of the required triangles is the piece BDE. The other three pieces may be rotated in the plane to form the second triangle.

—Aaron Buchman, *A.M.M.*, 58 (February, 1951), 112.

Indeed, the part cut off could be hinged at F and H.

Q 139. The Moving Digits

Let f denote the proper fraction which, when written as a decimal, consists of the repetend 15– – – –, one cycle of which is the number sought. According to the conditions of the problem, $5f = 0.----15$ $----15\cdots$ and $100f = 15.----15----15\cdots$, so that $95f = 15$, and $f = \frac{3}{19}$. By performing the division, the decimal repetend of f is found to consist of the 18-digit sequence, 157, 894, 736, 842, 105, 263. This is the unique number of less than 30 digits which satisfies the conditions of the problem. If the 30-digit limit were raised to 50, there would be just one more solution admitted, consisting of two cycles of f, or a number of 36 digits.

—H. T. R. Aude, *A.M.M.*, 41 (April, 1934), 268.

Q 140. Vertex of a Tetrahedron

If one face angle at one vertex of the tetrahedron is right or obtuse, the sum of the three face angles at this vertex exceeds π radians. If at least one face angle at each vertex is right or obtuse, the sum of the face angles at all of the vertices would exceed 4π. This is impossible since the sum of the interior angles of four triangles is just 4π radians. Consequently there must be at least one vertex of the tetrahedron at which all the face angles are acute.

—Roy MacKay, *A.M.M.*, 42 (August, 1935), 453.

Q 141. The Lucky Prisoners

The number of times t that the key was turned in the qth cell is equal to the number of divisors of q. Thus if $q = p_1{}^{a_1} \cdot p_2{}^{a_2} \cdots p_k{}^{a_k}$ where the p_i's are distinct primes, then $t = (a_1 + 1)(a_2 + 1) \cdots (a_k + 1)$. Now if any a_i is odd, t is even and the corresponding cell eventually remained locked. If all the a_i are even, q is a square number, t is odd, and the lucky occupants of the "square" cells found that their cells eventually remained open.

—*P.M.E.J.*, 1 (April, 1953), 330.

Q 142. No Point in Common

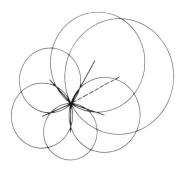

Assume, on the contrary, that the circular areas have a point in common.   Consider the six spokes connecting this point with each center.   At least one pair of spokes has an included angle not exceeding 60°.   The center associated with the shorter (or equal length) spoke then lies within the circular area whose center is associated with the other spoke.   This contradicts the assumption that none contains the center of any other.

—E. L. Magnuson, *A.M.M.*, 70 (May, 1963), 569.

Q 143. Mental Multiplication

Applying the identity $(a - b)(a + b) = a^2 - b^2$, we have

$$(96)(104) = (100 - 4)(100 + 4) = 10{,}000 - 16 = 9984.$$

Q 144. A Unique Triad

It is required to find distinct integer solutions of the set of simultaneous equations, $x + y = mz$, $y + z = nx$, $z + x = py$, where

m, n, p are positive integers. The condition that this set of homo-geneous equations has a solution is

$$\begin{vmatrix} 1 & 1 & -m \\ -n & 1 & 1 \\ 1 & -p & 1 \end{vmatrix} = 0.$$

That is, $mnp = m + n + p + 2$. Clearly, $(2, 2, 2)$ satisfies this equation. Any other solution must have one variable equal to 1. Whereupon $mn = m + n + 3$. Clearly, $(3, 3)$ satisfies this equation. Any other solution must have one variable less than 3. Thus $(5, 2)$ is the only other solution. Hence, there are but three triads of positive integers, $(2, 2, 2)$, $(1, 3, 3)$, and $(1, 2, 5)$, which satisfy the three-variable equation.

To each of these triads corresponds a set of three simultaneous equations in x, y, z. The solutions of these sets are (k, k, k), $(k, k, 2k)$, and $(2k, k, 3k)$, respectively. Therefore, the only set of distinct integers meeting the conditions of the problem is $(1, 2, 3)$.

$$-S.S.M., 49 \text{ (October, 1949)}, 590.$$

Q 145. The Enclosed Corner

The proposed floor space is a quadrilateral having one right angle and such that the two sides which meet in the vertex opposite the right angle are equal. Four congruent quadrilaterals of this form may be fitted together to form an equilateral octagon. For a given length of side, the octagon of maximum area is a regular octagon. Hence, one-fourth of this octagon will be the quadrilateral of maxi-mum area, $8(\sqrt{2} + 1)$ square feet. Therefore, the screens should be

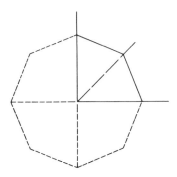

placed so that each forms with one wall and the bisector of the right angle at the corner of the room, an isosceles triangle of which the screen is the base.

—F. Hawthorne, *N.M.M.*, 19 (March, 1945), 322.

Q 146. Relatively Prime Integers

If $1/a + 1/b = 1/c$, then $a + b = ab/c$. Since a and b are integers, c must be composite, say $c = qr$, with one factor in common with a and the other factor in common with b. So let $a = mq$ and $b = pr$. Then $mq + pr = mqpr/qr = mp$. Since a, b, and c have no factor common to all three, m cannot divide r so it must divide p, and p cannot divide q so it must divide m. Hence $m = p$, so $p(q + r) = p^2$ and $q + r = p$. It follows that

$$a + b = pq + pr = p(q + r) = p^2,$$

$$a - c = pq - qr = q(p - r) = q^2,$$

and

$$b - c = pr - qr = r(p - q) = r^2.$$

—*S.S.M.*, 63 (October, 1963), 604.

Q 147. Diophantine Duo

The first equation may be written

$$(a - b - c)[a^2 + (b - c)^2 + ab + bc + ca] = 0.$$

Since the second factor cannot vanish for positive a, b, c, we must have $a = (b + c) = a^2/2$, whence the only solution in positive integers is $a = 2, b = c = 1$.

—E. W. Marchand, *A.M.M.*, 65 (January, 1958), 43.

Q 148. Bimedians of a Tetrahedron

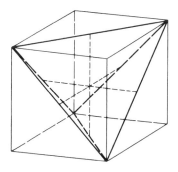

A regular tetrahedron may be inscribed in a cube with each pair of opposite edges of the tetrahedron coinciding with nonparallel diagonals of opposite faces of the cube. The midpoints of the edges of the tetrahedron are centroids of the faces of the cube. Hence, the join of the midpoints of two opposite edges of the tetrahedron passes through the centroid of the cube, is perpendicular to two opposite faces of the cube, and is parallel to four edges of the cube. Then since the three edges of a cube at a vertex are mutually perpendicular, the three joins of the midpoints of opposite edges of the regular tetrahedron are mutually perpendicular and are concurrent.

Q 149. Pied Product

Let $n = abc$ and let N be the desired product.   If $c = 1$, then the largest n, namely 981, gives $N = 159,080,922$ which is too small. Therefore $987 \geqq n \geqq 982$.   The sum of the digits of N is $35 \equiv 2$ (mod 3), so n and its permutations are congruent to 2 mod 3.   Then n is 986 or 983, of which only the latter has a product of digits ending in 6.   It follows that

$$(983)\,(839)\,(398) = 328,245,326.$$

—W. R. Talbot, *A.M.M.*, 66 (October, 1959), 726.

Q 150. A Peculiar Number

The common result must have 7 and 11 as factors, thus the number is $7 + 11$ or 18.   The method is general, since the solution of $(x - k)k = (x - m)m$ is $k + m$.

—*M.M.*, 33 (September, 1959), 58.

Q 151. Three of a Kind

It is assumed here that acquaintance between two persons is reciprocal.   If the six people are identified with the vertices of an octahedron, we have the equivalent problem: *If each edge and diagonal of an octahedron is colored red or green, then some triangle has all of its sides the same color.*

Each vertex is connected to every other vertex.   Of the five joins from one vertex, three must be of the same color.   If the ends of any

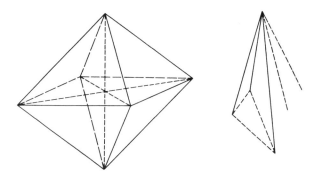

of these are joined by a line of the same color, there will be a triangle of that color. If the ends of each pair of the three are joined by a line of a different color, then these joins will form a triangle of that color.

Q 152. A Simplification Problem

Multiply numerator and denominator of the fraction by $2^{3/2}$ to secure

$$\frac{(8 + 2\sqrt{15})^{3/2} + (5 - 2\sqrt{15} + 3)^{3/2}}{(12 + 2\sqrt{35})^{3/2} - (7 - 2\sqrt{35} + 5)^{3/2}}$$

$$= \frac{(\sqrt{5} + \sqrt{3})^3 + (\sqrt{5} - \sqrt{3})^3}{(\sqrt{7} + \sqrt{5})^3 - (\sqrt{7} - \sqrt{5})^3} = \frac{2(5\sqrt{5} + 9\sqrt{5})}{2(21\sqrt{5} + 5\sqrt{5})} = \frac{7}{13}.$$

—S.S.M., 55 (October, 1955), 567.

Q 153. Product of Three Primes

Let $N = pqr$. Then $p^2 + q^2 + r^2 = 2331$, so no one of the primes exceeds $(2331)^{1/2} < 49$, and the primes are odd.

The sum of the divisors of N is $(1 + p)(1 + q)(1 + r) = 10{,}560 = 11 \cdot 960$. The only multiple of 11 less than 49 which is one more than a prime number is 44, so $r = 43$. Then $p^2 + q^2 = 482$ and neither exceeds $(482)^{1/2} < 22$. Now the squares of odd numbers end in 1, 5, or 9, so both p^2 and q^2 end in 1. Thus $p = 11$, $q = 19$, and $N = (11)(19)(43)$ or 8987.

Q 154. Representation of Rational Number

Let a/b be any positive rational number. Then $a/b = 1/b + 1/b + \cdots + 1/b$, a sum of harmonic terms with $a - 1$ duplications. Recursively expand all duplicate elements by the identity $1/n = 1/(n + 1) + 1/n(n + 1)$ until all terms are distinct.

—G. S. Cunningham, *A.M.M.*, 69 (May, 1962), 435.

For example: $\tfrac{3}{7} = \tfrac{1}{7} + \tfrac{1}{7} + \tfrac{1}{7} = \tfrac{1}{7} + \tfrac{1}{8} + \tfrac{1}{56} + \tfrac{1}{8} + \tfrac{1}{56}$

$= \tfrac{1}{7} + \tfrac{1}{8} + \tfrac{1}{56} + \tfrac{1}{9} + \tfrac{1}{72} + \tfrac{1}{57} + \tfrac{1}{3192}.$

Q 155. Condition That a Triangle Is Isosceles

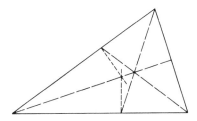

The bisector of an interior angle of a triangle divides the opposite side into segments proportional to the adjacent sides. If perpendiculars are dropped from a point to the sides of a triangle, then the sums of the squares of the alternate segments of the sides are equal.

Hence, in a triangle with sides a, b, c, if the perpendiculars to the sides at the feet of the internal angle bisectors are concurrent, then

$$\left(\frac{ab}{b+c}\right)^2 + \left(\frac{bc}{c+a}\right)^2 + \left(\frac{ca}{a+b}\right)^2$$

$$= \left(\frac{ca}{b+c}\right)^2 + \left(\frac{ab}{c+a}\right)^2 + \left(\frac{bc}{a+b}\right)^2.$$

Upon combining terms with the same denominators, we have

$$a^2(b-c)/(b+c) + b^2(c-a)/(c+a) + c^2(a-b)/(a+b) = 0,$$

and then

$$(b-c)(c-a)(a-b)(a+b+c)^2 = 0.$$

Since at least one of the first three factors must vanish, the triangle is isosceles.

—*A.M.M.*, 46 (October, 1939), 513.

Q 156. Classified Integers

If the integers are arranged in a square array as shown, each row is an arithmetic progression with a common difference of 1, and each column also is an arithmetic progression with $d = 4$. Conse-quently, the sum of any two numbers on the main

1	2	3	4
5	6	7	8
9	10	11	12
13	14	15	16

diagonals can be matched by at least one sum of a pair of numbers not on the diagonals. So the two sets are

A: 1   4   6   7   10   11   13   16,   and
B: 2   3   5   8   9   12   14   15.

—Wm. H. Benson

The 28 sums of pairs in each class are: 5, 7, 8, 10, 11(2), 12, 13, 14(2), 15, 16, 17(4), 18, 19, 20(2), 21, 22, 23(2), 24, 26, 27, 29. It will be observed that the sum of every pair-sum equidistant from the ends of this series is a constant, $34 = 2(1 + 16)$.

This result was obtained in another manner in $P.M.E.J.$, 3 (Spring, 1961), 182.

Q 157. Economical Ballots

Offhand, one would say $(3)(4)(5)$ or 60 different ballots.   However, if two fictitious names be added to the group of three, and one fictitious name to the group of four, then only five different ballots would be necessary.   Not only would this method reduce the printing costs, but it also would give statistics on whether or not members vote by relative order rather than by name.

—M. S. Klamkin, M.M., 30 (November, 1956), 110.

Q 158. Comparison of Ratios

We have $(x^2 - y^2)/(x - y) = x + y > x + y - 2xy/(x + y) = (x^2 + y^2)/(x + y)$.

Q 159. Mixtilinear Triangle

The centers of the inscribed circle and the given semicircle are at (r, r) and $(a/2, b/2)$, respectively, on a coordinate system defined by the legs of the given triangle, where r is the unknown radius. Consequently, the distance between the centers is equal to the difference in the radii.   That is, $(r - a/2)^2 + (r - b/2)^2 = (c/2 - r)^2$.

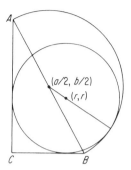

Whereupon, noting that $a^2 + b^2 = c^2$, we have $r = a + b - c$, which also is the diameter of the inscribed circle of the right triangle.

—M. A. Kirchberg, *A.M.M.*, 62 (June, 1955), 444.

Q 160. Pandiagonal Heterosquare

In summing rows, columns, and diagonals of such an array, each integer would be counted four times. The sum of all such sums is therefore $4[n^2(n^2 + 1)/2]$ or $2n^2(n^2 + 1)$. If there were a least integral sum k such that the $4n$ sums ranged from k to $k + 4n - 1$ consecutively, the grand total also would be given by $2n(2k + 4n - 1)$. Equating these two expressions and simplifying we have $n(n^2 + 1) = 2k + 4n - 1$. But the left side is always even whereas the right side is always odd. Hence, k cannot exist, nor n such that the $4n$ sums are consecutive numbers.

—D. C. B. Marsh, *A.M.M.*, 62 (January, 1955), 42.

Q 161. A Product of 2^{m+1}

Consider the expression $I = (\sqrt{3} + 1)^{2m} + (\sqrt{3} - 1)^{2m}$. This obviously is an integer, and since $(\sqrt{3} - 1)^{2m}$ is less than unity, I is the integer next greater than $(\sqrt{3} + 1)^{2m}$.   Then

$$I = (4 + 2\sqrt{3})^m + (4 - 2\sqrt{3})^m = 2^m[(2 + \sqrt{3})^m + (2 - \sqrt{3})^m]$$

$$= 2^{m+1}[2^m + 2^{m-2}(3)m(m - 1)/2 + \cdots].$$

—Cecil B. Read, *S.M.M.*, 62 (November, 1962), 622.

Q 162. In a Nonagon

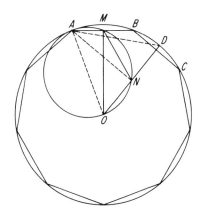

Arc $AB = 40°$, then arc $AD = 60°$, so triangle AOD is equilateral and AN is perpendicular to OD.   Consequently, A, M, N, O are concyclic on a circle with AO as diameter.   Hence angle $OMN =$ angle $OAN = 30°$, since they are inscribed in the same arc.

—Howard Eves, *A.M.M.*, 63 (June, 1956), 423.

Q 163. Spending Money

Twenty-four quarters are involved. $24 = (2)(12) = (3)(8) = (4)(6)$.   From these factorizations we pick two such that $q_1c_1 = q_2c_2$

and $q_1 + 1 = q_2$ while $c_1 = c_2 + 2$.   Thus there were eight children in the final party.

—M.M., 31 (January, 1958), 217.

Q 164. Summing an Infinite Series

Let

$$S = 1 + 2x + 3x^2 + 4x^3 + \cdots$$
$$xS = \quad\; x + 2x^2 + 3x^3 + \cdots$$
$$\overline{(1 - x)S = 1 + \; x + \; x^2 + \; x^3 + \cdots = 1/(1 - x)}$$
$$S = 1/(1 - x)^2.$$

Q 165. A Fenced Square Field

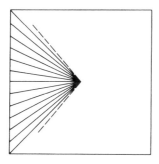

Divide the square field into triangles each bounded by one chain (66 feet) of fence and two radial lines to the center of the square. These triangles have equal areas and altitudes equal to one-half the side x of the square.   Thus each triangle has as many boards as acres.   That is, (6)(4) boards and 24 acres or 240 square chains.

Hence $(\frac{1}{2})\,(x/2)\,(1)\; =\; 240$, so the side of the field is 960 chains or 12 miles.

Indeed, if the field were any regular polygon (or a circle) with 11-foot boards bent around the corners (or arcs), the area would be $(\frac{1}{2})\,$(perimeter)(radius of the inscribed circle), so $4p/11\; =\; (\frac{1}{2})\,pr/43{,}560$ and $2r\; =\; 63{,}360$ feet or 12 miles.

—R. R. Rowe, *Civil Engineering—ASCE*, 11 (January, 1941), 70.

Q 166. Triangular Numbers from Odd Squares

We have $(2n+1)^2 = 4n^2 + 4n + 1 = 4n(n+1) + 1 = 8k + 1$, since of two consecutive integers, one is even. To cut off the last digit is equivalent to dividing the remaining $4n(n+1)$ by the base eight to produce $n(n+1)/2$, a triangular number.

—G. W. Wishard, *N.M.M.*, 10 (May, 1936), 313.

Q 167. Inscribed Circles

Neither.   The inradius of a triangle is given by

$$r = [(s-a)(s-b)(s-c)/s]^{1/2} \quad \text{where} \quad 2s = a + b + c.$$

For each of the triangles in this problem the inradius is found to be 6.

A rare case of obtuse triangle twins is 97, 169, 122 and 97, 169, 228, each of which has an inradius of 30.

—Bancroft H. Brown, *M.M.*, 29 (May, 1956), 275–276.

Q 168. An Inscribed Dodecagon

Let AB and BC be a pair of adjacent but unequal sides.   Then arc $AB +$ arc $BC =$ arc $AC = 360°/6 = 60°$, so $AC = r$.   Applying

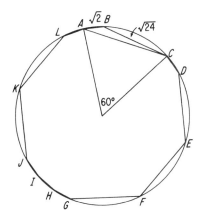

the law of cosines to triangle ABC, wherein inscribed angle ABC = $150°$,

$$(AC)^2 = (\sqrt{2})^2 + (\sqrt{24})^2 - 2\sqrt{2}\sqrt{24}(-\sqrt{3}/2) = 38.$$

Hence, $r = \sqrt{38}$.

—Norman Anning, *M.M.*, 28 (November, 1954), 113.

Q 169. A Susceptible Diophantine Equation

Since $2^{24} + 2^{24} = 2^{25}$, we have $(2^8)^3 + (2^6)^4 = (2^5)^5$ or $a = 256$, $b = 64$, $c = 32$.

—Leo Moser, *M.M.*, 26 (September, 1952), 53.

Q 170. Antifreeze

One quart of the old solution differs from one quart of the new (or average) solution by -24 percent, while one quart of the solution to be added differs from the new solution by $+48$ percent. Hence, there must be two quarts of the old solution for each quart of the

155

added solution.   So $\frac{1}{3}$ of the original radiator content or 7 quarts must be drained.

—B. E. Mitchell, *M.M.*, 26 (January, 1953), 153.

Q 171. Maximum-minimum without Calculus

Let

$$f(x) = \frac{x^2 - 2x + 2}{2x - 2} = \frac{1}{2}\left[x - 1 + \frac{1}{x - 1}\right] = \frac{x^2}{2(x - 1)} - 1.$$

The smallest absolute values of the sum of a number and its reciprocal are attained when the number is ± 1.   Hence the smallest absolute values of $f(x)$ will be given by $x - 1 = \pm 1$, so $x = 2$ or 0.   From the last form of $f(x)$, it is evident that $f(x)$ will increase without bound for $x > 2$, will increase negatively without bound for $x < 0$, and is undefined for $x = 1$.

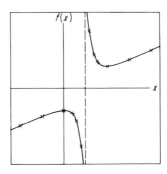

It follows that for $x = 2$, $f(x)$ has a relative minimum value of 1, and for $x = 0$, $f(x)$ has a relative maximum value of -1.

Q 172. Tangent Sum Equal to Product

We have $360° - 125° = 117° + 118°$. So,

$$-\tan 125° = \tan(117° + 118°)$$
$$= (\tan 117° + \tan 118°)/(1 - \tan 117° \tan 118°).$$

Simplifying,

$$\tan 117° \tan 118° \tan 125° = \tan 117° + \tan 118° + \tan 125°.$$

—Norman Anning, *M.M.*, 32 (November, 1958), 113.

In general, the sum of the tangents of A, B, and $(360° - A - B)$ equals their product.

Q 173. A Faded Document

The legible digits establish that $3 + 1 = 10$, so the computation is in the scale of four, in which there are only three nonzero digits. One of these multiplying the divisor must give **1, and another, 3*. Now $3(13) = 111$, $3(23) = 201$, $3(33) = 231$, and $2(13) = 32$, $1(33) = 33$. But $2(13) + 1 < 100$, so on the basis of the first subtraction, the divisor must be 33.   It follows that the quotient is 1031 and the dividend is 102,003.

Q 174. Bisecting Yin and Yang

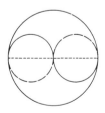

About the diameter, revolve the configuration 180° onto itself.

Each of the two circles thus formed has $\frac{1}{2}$ the diameter and hence $\frac{1}{4}$ the area of the large circle.   So each of the other portions of the large circle also is $\frac{1}{4}$ the area of the large circle.

—*M.M.*, 34 (November, 1960), 107–108.

Q 175. Number That Is Factor of Its Reverse

In any system of numeration with base $B \geqq 10$,

$$2(297) < 2(300) = 600 < 792 < 800 = 4(200) < 4(297).$$

Hence, $792 = 3(297)$, so that $7B^2 + 9B + 2 = 3(2B^2 + 9B + 7)$ or $B^2 - 18B - 19 = 0$.   Rejecting the negative root leaves $B = 19$.

—D. L. Silverman, *M.M.*, 38 (March, 1965), 124.

Q 176. Square Dad

The legality of the marriage reduces the possibilities to three and again to one, the middle possibility below, when the ages of the children are considered.

64	10		49	13		36	9
24	6		36	9		18	9

So the ages of the father, mother, daughter, and son are 49, 36, 13, and 9, respectively.   Although $1 + 3 = 4$ and there are four members of the family, it just misses being square throughout.   But, mother is a perfect 36.   It just happens that the sum of the daughter's and mother's ages equals the father's age.

Q 177. Tetrahedron through a Straw

In a parallelogram consisting of a strip of four equilateral triangles, lines drawn parallel to a long side have a constant length $2e$.   When

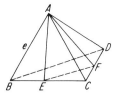
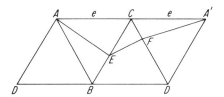

the strip is folded into a regular tetrahedron, it follows that the sections of the tetrahedron made by planes perpendicular to the join of the midpoints of two opposite edges have a constant perimeter $2e$.

Consequently, when its bimedian coincides with the axis of the cylinder, the tetrahedron may be pushed through a flexible thin-walled cylinder with a circumference $\pi d = 2e$. Thus $e = \pi d/2$. In practice, it would be helpful to have the end of the cylinder flared out slightly in order to get the job underway.

If the tetrahedron has a different attitude to the axis of the cylinder, some plane perpendicular to the axis will pass through a vertex and cut two edges not issuing from that vertex. As can be seen from the developed surface in the figure, the perimeter of a typical section $AEFA'$ is greater than $2e$. Consequently, the tetrahedron cannot pass through the cylinder in this attitude. It follows that the largest tetrahedron that can pass through the cylinder is one with edge $\pi d/2$.

—*M.M.*, 39 (March, 1966), 133.

Q 178. A Product of $(a - 1)^2$

$$f(a, n) = a^{n+1} - n(a - 1) - a = a(a^n - 1) - n(a - 1)$$

$$= (a - 1)[a(a^{n-1} + a^{n-2} + \cdots + 1) - n].$$

Since the expression in the brackets vanishes for $a = 1$, then by the factor theorem, the expression is divisible by $a - 1$, so $(a - 1)^2$ divides $f(a, n)$.

Q 179. Determinant of Pascal's Triangle

By the law of formation, each element in the array is equal to the sum of the element immediately above it and the element to its left. Hence in an nth-order determinant based on the first row, performance of the operations $\text{col}_i - \text{col}_{i-1}$, $i = n$, $(n - 1)$, $\cdots$, 2 will reduce the determinant to the minor of its lower left element. Expansion by minors of the elements of the first row, and continuous repetition of the sequence of operations on the derived determinants ultimately results in the value 1. For example:

$$
\begin{vmatrix} 1 & 1 & 1 & 1 \\ 3 & 4 & 5 & 6 \\ 6 & 10 & 15 & 21 \\ 10 & 20 & 35 & 56 \end{vmatrix}
=
\begin{vmatrix} 1 & 0 & 0 & 0 \\ 3 & 1 & 1 & 1 \\ 6 & 4 & 5 & 6 \\ 10 & 10 & 15 & 21 \end{vmatrix}
$$

$$
=
\begin{vmatrix} 1 & 1 & 1 \\ 4 & 5 & 6 \\ 10 & 15 & 21 \end{vmatrix}
=
\begin{vmatrix} 1 & 0 & 0 \\ 4 & 1 & 1 \\ 10 & 5 & 6 \end{vmatrix}
=
\begin{vmatrix} 1 & 0 \\ 5 & 1 \end{vmatrix}
= 1.
$$

Since the array is symmetrical about its principal diagonal, the proof also applies to determinants of arrays based on the first column. In this case the rows rather than the columns would be subtracted.

Q 180. Rational Coordinates

The given cubic consists of a straight line and an ellipse, as may be seen from the equivalent form of its equation

$$
(x + y - 1)(2x^2 - 2xy + 2y^2 - x - y - 1) = 0.
$$

From the first factor it is clear that if any rational value be assigned to x, then at least one rational value of $y = 1 - x$ can be computed.

Q 181. A Closed Construction

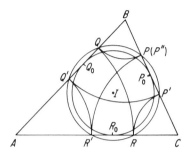

Let P_o, Q_o, and R_o be the points of contact of the incircle.   Then $PP_o = QQ_o = RR_o = P'P_o = Q'Q_o = R'R_o = P''P_o$.   This shows that the construction closes, and that P, Q, R, P', Q', R' lie on a circle concentric with the incircle.

—Howard Eves, *A.M.M.*, 50 (June, 1943), 391.

This generalizes to any odd polygon possessing an incircle.   Also, the construction will close if the odd polygon has sides which may be made to touch a circle by changing the angles but not the length of the sides.

Q 182. Grouped Odd Integers

The nth group contains n integers, so the number of integers in all the groups including the nth is $n(n + 1)/2$, and the number of integers in all the groups previous to the nth is $(n - 1)n/2$.   These two sets are arithmetic progressions with a common difference of 2.

Hence the sum of the members of the nth group is

$$\frac{1}{2}\frac{n(n+1)}{2}\left\{2+\left[\frac{n(n+1)}{2}-1\right]2\right\}$$

$$-\frac{1}{2}\frac{(n-1)n}{2}\left\{2+\left[\frac{(n-1)n}{2}-1\right]2\right\}$$

$$=n^2[(n+1)^2-(n-1)^2]/4=n^3.$$

Q 183. The Sixteen-point Sphere

Let R be the circumradius and r the radius of the sixteen-point sphere. If the tetrahedron is regular, then $r = R/3$; and if it is trirectangular, $r = \infty$ (since the circumcenters of the faces are coplanar) with R finite. Since the regular tetrahedron may be continuously deformed into a trirectangular one, it follows that at some intermediate stage we must have $r = R/2$. Indeed, any value of $r/R > \frac{1}{3}$ is possible.

—Howard Eves, *A.M.M.*, 50 (June, 1943), 389.

Q 184. Concurrent Circles

Let the circles with centers at O_1, O_2, O_3 be concurrent at A, and the pair (O_1), (O_2) also at B, (O_1), (O_3) at C, and (O_2), (O_3) at D. O_3O_1 and O_3O_2 are perpendicular to CA and AD, respectively, so angle $O_1O_3O_2$ + angle $CAD = 180°$. In circle (O_1), angle $ACB = \frac{1}{2}$ arc AB = angle AO_1O_2 (since O_1O_2 is the perpendicular bisector of AB). In (O_2), angle $ADB = \frac{1}{2}$ arc AB = angle AO_2O_1. Now

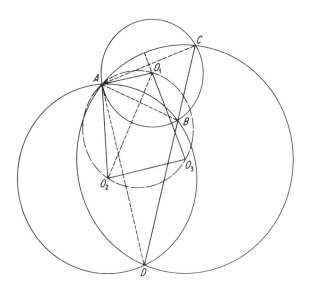

C, B, D are collinear, so triangles ACD and AO_1O_2 are similar, and angle CAD = angle O_1AO_2. It follows that angle $O_1O_3O_2$ + angle O_1AO_2 = 180°, and $O_1O_3O_2A$ is inscriptible in a circle.

$$—A.M.M., 72 \text{ (May, 1965), 547.}$$

Q 185. Golf Tournament

Let certain permutations of four letters be designated as follows: $P_1(xyzw)$, $P_2(zwxy)$, $P_3(wzyx)$, and $P_4(yxwz)$.

Each man plays once with each of the other 15, so there must be $15\!/\!3$ or 5 rounds. One way to set up the members in foursomes is to first arrange the initials of the 16 names in a square array. Then generate three more arrays by repeating the first column, then apply P_4, P_3, P_2, successively to the second column, then P_2, P_4, P_3 to the third column, and finally P_3, P_2, P_4 to the fourth column. Thus

```
A E I M     A F K P     A H J O     A G L N
B F J N     B E L O     B G I P     B H K M
C G K O     C H I N     C F L M     C E J P
D H L P     D G J M     D E K N     D F I O
```

The foursomes for the first four rounds are given by the rows of the four arrays, and the fifth round by the columns of the first array. The rows in the last three arrays constitute the positive terms of the expansion of the first array as a determinant.

The negative terms of the expansion are obtained by applying P_3, P_4, P_2 in order to the second column, then P_2, P_3, P_4 to the third column, and then P_4, P_2, P_3 to the fourth column of the first array. Thus a different scheduling is obtained:

```
A E I M     A H K N     A F L O     A G J P
B F J N     B G L M     B E K P     B H I O
C G K O     C F I P     C G J M     C E L N
D H L P     D E J O     D H I N     D F K M
```

Q 186. Square Triangular Numbers

If $T[n] = n(n + 1)/2$, the nth triangular number, is a square, then $T[4n(n + 1)] = 4T[n](2n + 1)^2$ is also a square. Since the first triangular number, 1, is also a square, there exist an infinite number of square triangular numbers.

—A. V. Sylwester, *A.M.M.*, 69 (February, 1962), 168.

Q 187. System in Three Variables

In consideration of the relation between the coefficients and the roots of an equation, evidently x, y, z are the roots of the cubic equation

$$a^3 - 6a^2 + 11a - 6 = 0.$$

Then $$(a - 1)(a - 2)(a - 3) = 0.$$

$$a = 1, 2, 3.$$

Since the equations are symmetrical in x, y, z the six solutions are the permutations of 1, 2, 3, namely:

$$(x, y, z) = (1, 2, 3), (1, 3, 2), (2, 1, 3), (2, 3, 1), (3, 1, 2), (3, 2, 1).$$

—D. G. Buckley, *S.S.M.*, 40 (May, 1940), 483.

Q 188. Imbedded Polyhedron

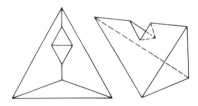

It can be a tetrahedron with a notch in the form of a tetrahedral wedge removed from one edge. This is most easily seen by redrawing the graph with one of the triangles outermost, or by viewing the notched tetrahedron through one of the remaining triangular faces.

—Robert Connelly, *A.M.M.*, 69 (December, 1962), 1009.

Q 189. Baseball Team Standings

The basic procedure relates the $W{-}L$ record of a team to a 50–50 situation. If its $W > L$, the team is in the upper bracket. If its $W < L$, the team is in a lower bracket. Now consider two teams A and B. Team A will be above B if $W_A > W_B$ and $L_A \leqq L_B$ or $W_A = W_B$ and $L_A < L_B$. Now we can order the teams, with doubts only about the relative placement in the marked pairs. Thus

	W	L			W	L
*Cincinnati	49	36		St. Louis	41	45
*Los Angeles	51	38		*Chicago	41	46
San Francisco	45	38		*Houston	39	45
Philadelphia	45	39		New York	29	56
*Milwaukee	42	40				
*Pittsburgh	44	43				

If $W_A = W_B + x$ and $L_A = L_B + y$, then if both are in the upper bracket and $x \leqq y$, A is below B. If both are in the lower bracket and $x \geqq y$, A is above B. So the order above is the correct one.

Q 190. A Sweet Purchase

Since $216 = 2^3 \cdot 3^3 = (8)(27) = (9)(24)$, she bought 24 pounds.

—M.M., 33 (November, 1959), 118.

Q 191. Intersections of Diagonals

Consider a convex polygon of $n \geqq 4$ sides. Every combination of the n vertices taken four at a time determines a quadrilateral which

has two intersecting diagonals.   Also, every two intersecting diagonals of the polygon determine a quadrilateral.   Therefore, the required number of intersections is $C(n, 4)$.   All these intersections, of course, may not be distinct.

—Norbert Kaufman and R. H. Koch, *A.M.M.*, 54 (June, 1947), 344.

Q 192. Two Vanishing Triads

Given $a + b + c = 0$ and $d + e + f = 0$.   Then

$$(a + b)^3 = (-c)^3$$

$$a^3 + b^3 + c^3 = -3ab(a + b) = 3abc.$$

In like manner,

$$d^3 + e^3 + f^3 = -3de(d + e) = 3def.$$

Finally,

$$(a^3 + b^3 + c^3)/(d^3 + e^3 + f^3) = abc/def.$$

—Aaron Buchman, *S.S.M.*, 38 (February, 1938), 220.

Q 193. Condition for Factorability

Let $f(x) = x^a + x^b + 1 = x^a + x^{3k}x^{-a} + 1$.   Now the cube roots of unity are 1, $\omega = (-1 + i\sqrt{3})/2$, and ω^2.   Since a is prime to 3, then either $\omega^a = \omega$ and $\omega^{-a} = \omega^2$, or $\omega^a = \omega^2$ and $\omega^{-a} = \omega$.   In either case,

$$f(\omega) = f(\omega^2) = 1 + \omega + \omega^2 = 0.$$

Therefore, $(x - \omega)(x - \omega^2) = x^2 + x + 1$ is a factor of $x^a + x^b + 1$.

—Howard D. Grossman, *S.S.M.*, 45 (May, 1945), 486.

Q 194. Parallel Resistances

Since x, y, z are positive, $x > z$ and $y > z$.   Let $x = z + u$ and $y = z + v$, $u, v > 0$.   The equation $1/z = 1/x + 1/y$ reduces to $z^2 = uv$.   Thus for every z, we merely have to decompose z^2 in every way as a product of two integers u and v.

—Marion L. Gaines, *N.M.M.*, 19 (November, 1944), 100.

Q 195. Minimum Bisector

Reflect the triangle and the bisecting curve repeatedly, keeping one vertex fixed, and thus produce a regular hexagon.   The closed curve cuts the area of the hexagon in half, so has a fixed area inside. Consequently, if its perimeter is a minimum, the curve is a circle with center at the fixed vertex.

—Leo Moser.

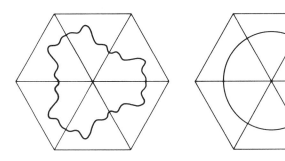

The arc of the circle bisecting the triangle is $< a/\sqrt{2}$, where a is the side of the triangle and $a/\sqrt{2}$ is a bisecting line parallel to the base.

Q 196. A Peculiar Square

We have

$$N^2 = a_1a_2a_3b_1b_2b_3a_1a_2a_3 = (a_1a_2a_3)(1002001) = (a_1a_2a_3)(7^2)(11^2)(13^2).$$

Therefore, $a_1a_2a_3$ must be the square of a prime P other than 7, 11, or 13.   Now $a_1 \neq 0$ and $b_1b_2b_3 < 1000$, so $10 < P < 23$.   Hence, P is 17 or 19, $a_1a_2a_3 = 289$ or 361, and $N^2 = 289{,}578{,}289$ or $361{,}722{,}361$.

—P. N. Nagara, *M.M.*, 24 (November, 1950), 108.

Q 197. A Series of Tests

If a difference in grade of $97 - 73$ or 24 points would change an average grade $90 - 87$ or 3 points, there must have been $2\frac{2}{3}$ or 8 tests.

Q 198. Coinciding Points in a Quadrilateral

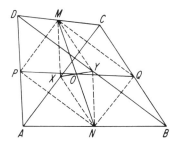

The midpoints of the sides of any quadrilateral are the vertices of a parallelogram, and the diagonals of a parallelogram bisect each other. In the quadrilateral $ABCD$, $MQNP$ is a parallelogram, so PQ and MN intersect at O, the midpoint of each. In the crossed quadrilateral $ACDB$, $MYNX$ is a parallelogram, so XY and MN intersect at O, the midpoint of each.

—Louis R. Chase, *S.S.M.*, 31 (May, 1931), 616.

Q 199. Fractions in Lowest Terms

If k/n is a fraction in lowest terms, then $1 - k/n$ or $(n - k)/n$ is also. Thus fractions in lowest terms can be arranged in matched pairs, so the number of terms is even.

—Norman Anning, *M.M.*, 27 (May, 1954), 284.

Q 200. The Tea Set

Since the cost is less than the retail price, it is evident from the prices of the sugar bowl and the creamer that $H = 0$. Then

$$KOC/CKO = 672/600 = 28/25.$$

It follows that KO is a multiple of 25, so $K = 5$. Also, C is even and less than K, so $KOC = 504$ and $CKO = 450$. Hence, the cost is $^{450}\!/_{600}$ or $\frac{3}{4}$ of the retail price. Immediately, the tray cost \$37.62, the teapot cost \$68.31, T must be 9, and the key word of the code is

$$1\ 2\ 3\ 4\ 5\ 6\ 7\ 8\ 9\ 0$$
$$B\ L\ A\ C\ K\ S\ M\ I\ T\ H$$

Q 201. Segments Determining an Equilateral Triangle

In the triangle ABC, $PC:PA:PB::3:4:5$.
Construct an equilateral triangle PCF so that P and F are on

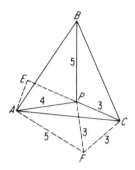

opposite sides of AC. Draw AF. Drop AE perpendicular to CP extended. $\angle PCB = 60° - \angle PCA = \angle ACF$, so triangles PCB and FCA are congruent, and $AF = BP = 5$. Hence, APF is a right triangle, so $\angle APE = 180° - 60° - 90° = 30°$.   It follows that $AE = 2$ and $EP = 2\sqrt{3}$.   Then

$$AC = \sqrt{(2)^2 + (3 + 2\sqrt{3})^2} = \sqrt{25 + 12\sqrt{3}} \doteq 6.7664 \text{ inches.}$$

—S.S.M., 33 (April, 1933), 450.

Q 202. An Invariant Remainder

If the remainder is to be the same, the divisor must be odd.   Now $1453 - 1108 = 345$, $1844 - 1453 = 391$, $2281 - 1844 = 437$. Then $437 - 391 = 391 - 345 = 46 = 2(23)$.   So 23 is the required divisor, since

$$(N_1 d + r) - (N_2 d + r) = d(N_1 - N_2).$$

The remainder is 4.

—M.M., 35 (January, 1962), 62.

Q 203. Nine Non-Zero Digits

The only possible values of c are 3 and 4. Now $(64644)^2$ has ten digits, so $(27,273)^2 = 743,816,529$ is the unique solution.

—Nick Farnum, *S.S.M.*, 63 (October, 1963), 603.

The solution is unique even without the $ab = c^3$ restriction. Furthermore, $27,273 = 3(9091)$, and 9091 is the largest prime factor of any square number composed of the nine positive digits used once each.

Q 204. Dissection for Coincidence

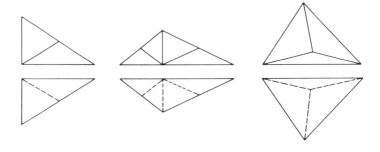

The result can be achieved by dissecting one of the triangles into isosceles triangles. A right triangle will be divided by the median upon the hypotenuse. An obtuse triangle can first be divided into right triangles by the altitude upon the longest side. An acute triangle will be divided into three isosceles triangles by the circumradii to the vertices.

—Louis R. Chase, *S.S.M.*, 30 (November, 1930), 949.

Q 205. Deflating Umbugio

Let $b = (x - 1/x)^{1/2}$ and $a = (1 - 1/x)^{1/2}$. Then the original equation becomes

$$x = a + b \qquad (1)$$

and since $x \neq 0$,

$$b - a = (b^2 - a^2)/(b + a) = (x - 1)/x = 1 - 1/x \qquad (2)$$

Adding (1) and (2) we find

$$2b = x - 1/x + 1 = b^2 + 1, \quad \text{so} \quad b = 1.$$

Whereupon, $x - 1/x = 1$, $x^2 - x - 1 = 0$, and $x = (1 \pm \sqrt{5})/2$. However, the only value satisfying the original equation is $x = (1 + \sqrt{5})/2$.

—P. M. Anselone and Sam Cook, *A.M.M.*, **62** (December, 1955), 728.

Q 206. Males with Common Characteristics

One hundred of the males would have $70 + 75 + 85 + 90$ or 320 of the recorded characteristics. In the most uniform distribution each male would have three of the characteristics, so at least twenty, that is, 20 percent would have all four characteristics. In general,

$$P = \sum_{i=1}^{n} p_i - 100(n - 1).$$

—*M.M.*, **38** (September, 1965), 211.

Q 207. An Area of Constant Width

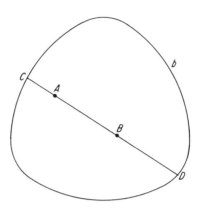

Let C and D be the points where the line AB meets the boundary b. Since the region has a constant width (diameter) 1, then $CD \leqq 1$. But $(AC + CB) + (BD + DA) = 2\,CD$, so one of $(AC + CB)$ and $(BD + DA)$ must be $\leqq 1$.

—Leo Moser.

Q 208. Comparing Radicals

In general, $n! < (n + 1)^n$, since each one of the n factors on the left is less than $n + 1$. Then

$$(n!)^n n! < (n!)^n (n + 1)^n \quad \text{or} \quad (n!)^{n+1} < [(n + 1)!]^n.$$

Extracting the $n(n + 1)$th root, $(n!)^{1/n} < [(n + 1)!]^{1/(n+1)}$.

Or, in the specific case given, assume that $(8!)^{1/8} < (9!)^{1/9}$. Raise both sides to the 72d power and obtain $(8!)^9 < (9!)^8$. Dividing by $(8!)^8$, we have $8! < 9^8$ which obviously is true, so the assumption is correct.

Q 209. Fibonacci Tetrahedron

The general formula for the Fibonacci series is $F_n + F_{n+1} = F_{n+2}$. Hence, any three consecutive Fibonacci numbers satisfy the equation, $x + y = z$. Consequently, the four vertices of the tetrahedron are coplanar and the volume is zero.

Indeed, the coordinates need not be 12 consecutive terms. The same situation exists for any four triads of consecutive Fibonacci numbers.

Q 210. The Regular Octahedron

(a) A regular octahedron may be viewed as an antiprism, with the parallel plane cutting the six lateral equilateral triangular faces. By cutting along an edge, say AD, the six lateral faces can be flattened into a parallelogram $ADD'A'$. Hence the perimeter of the section is equal to the perimeter of the triangular base of the octahedron, namely $3e$.

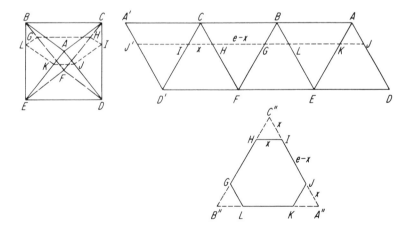

(b) If $A J = x$, then the section is a hexagon made by cutting off equilateral triangles of side x from the corners of an equilateral triangle of side $e + x$. Hence the hexagon has sides which are alternately x and $e - x$. Its area is $[(e + x)^2 - 3x^2]\sqrt{3}/4$ which will be a minimum when $x = 0$ and a maximum when $x = e/2$. The maximum hexagon is a regular hexagon with vertices at the midpoints of the lateral edges of the octahedron. Its area is $\frac{3}{2}$ the area of a face of the octahedron.

Q 211. Never an Integer

In the sequence 1, 2, 3, $\cdots$, n there can be only one term θ which contains the highest power of 2. Let the least common multiple of the sequence be $2M$. Now multiply both sides of the equation, $S = 1 + \frac{1}{2} + \frac{1}{3} + \cdots + 1/n$ by M. Then each term of the right member will be an integer except M/θ which will have a denominator of 2. Hence SM is not an integer, nor is S.

Q 212. Three Consecutive Odd Integers

All integers are of one of the forms $4k$, $4k + 1$, $4k + 2$, or $4k + 3$. It follows that their squares are of the forms $4k$ or $4k + 1$, and the sum of two squares is of the form $4k$, $4k + 1$, or $4k + 2$. Now every other odd integer is of the form $4k + 3$, so not even each of *two* consecutive odd integers can be the sum of two squares.

Q 213. Vanishing Vector Sum

Let R be the resultant of the vectors. About O, rotate the configuration through $2\pi/n$ radians, bringing it into coincidence with its former position.　Note that R also rotates through $2\pi/n$ radians to become R'.　Clearly, $R = R'$, but since their directions differ, then $R = R' = 0$.

—Richard Couchman, M.M., 26 (May, 1953), 287.

Q 214. The Sliding Ellipse

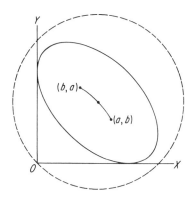

The locus of the intersection of two perpendicular tangents to the ellipse $b^2x^2 + a^2y^2 = a^2b^2$ is the director circle $x^2 + y^2 = a^2 + b^2$. Hence, the center of the moving ellipse which is tangent to two fixed perpendicular lines, considered as axes of reference, is always the distance $(a^2 + b^2)^{1/2}$ from the origin.　The closest approach of the center to the axes is at (a, b) and (b, a), so the locus of the center is the minor arc of the circle $x^2 + y^2 = a^2 + b^2$ included between these points.　[The length of the arc is $(a^2 + b^2)^{1/2} \arctan (a^2 - b^2)/2ab$.]

—Adrian Struyk, M.M., 24 (March, 1951), 231.

Q 215. Superposed Radical

A general term of the sequence 11, 14, 17, $\cdots$ is $8 + 3n$. A general term of the sequence 4, 10, 18, $\cdots$ is $n^2 + 3n$. The first integer cube greater than 11 is 27 or 3^3. To get 27 under the first radical sign there has to be $4^3 = 64$ under the second one, since $11 + 16 = 27$. But with $64 = 14 + 50$ under the second radical sign, $5^3 = 125$ must be under the third one. In general, under the $(n + 1)$st radical sign there must be $(n + 3)^3$, in order that there may be $(8 + 3n) + (n^2 + 3n)\sqrt[3]{(n + 3)^3}$ or $(n + 2)^3$ under the nth radical sign. Therefore, the value of the given expression is 3.

—Edgar Karst, *M.M.*, 32 (January, 1959), 169.

Q 216. Fibonacci Pythagorean Triangles

In the Fibonacci sequence, $F_{n+2} = F_n + F_{n+1}$, $F_1 = F_2 = 1$. If three numbers chosen from the sequence, in order of magnitude are a, b, c, then $c \geqq a + b$. Hence, they cannot represent the sides of *any* triangle, since in a triangle the sum of two sides is always greater than the third side.

—Norman Miller, *A.M.M.*, 60 (March, 1953), 191.

Q 217. Hole in Sphere

Consider a sphere of radius k with an expansible membrane surface which acts to contain the fluid interior according to the laws of surface tension. Now pierce the sphere along a diameter and insert a radially expansible tube of constant length $2k$ in such a way as to lose no fluid. As the tube expands, the surface tension will maintain the membrane in a spherical contour of increasing radius R—but the

 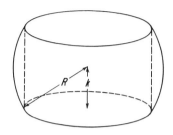

volume of the "wedding ring" will remain constant. Though the interior circumference of the "ring" or pierced sphere, $2\pi\sqrt{R^2 - k^2}$, will increase, its equatorial thickness, $R - \sqrt{R^2 - k^2}$, will decrease. Thus the volume of the material remaining after a sphere of radius R is pierced by a hole of length $2k$ is the same as the volume of a sphere of radius k, that is, $4\pi k^3/3$. Hence the remaining volume is independent of the radius of the sphere. Specifically, $V = 4\pi(5^3)/3 \doteq 523.6$ cubic inches.

Q 218. The Commuter

Two round trips made the first way would take 3 hours, thus covering the distance between home and office twice walking and twice riding. Therefore, he could make the round trip by walking in $3 - \frac{1}{2}$ or $2\frac{1}{2}$ hours.

Q 219. Odd Base of Notation

Any integer in the scale of r may be represented as $a_0 r^n + a_1 r^{n-1} + \cdots + a_n r^0$. If a product has an even factor it is even, otherwise it is odd. Now if r is odd, r^k is odd, and the characters of the members of the indicated sum are determined by the characters of the a's.

Now the sum of even numbers is even, as is the sum of an even number of odd numbers. The sum of even numbers and an odd number of odd numbers is odd. Hence, an integer in an odd scale is odd, if and only if it has an odd number of odd digits.

—*N.M.M.*, 12 (January, 1938), 197.

Q 220. Journey on a Dodecahedron

The rhombic dodecahedron has eight three-edged vertices T and six four-edged vertices F. The neighbors of each T are F's and the neighbors of each F are T's. Hence for each journey the T's and F's have to alternate. But eight T's and six F's cannot be written as one alternating sequence, whether we want to return to the starting point or not.

—Arthur Rosenthal, *A.M.M.*, 53 (December, 1946), 593.

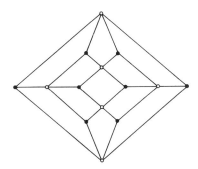

The T's are the vertices of a cube and the F's are vertices of a regular octahedron. The impossibility of the journey can be more easily visualized with the aid of the Schlegel diagram from H. S. M. Coxeter, *Regular Polytopes*, Methuen (1948), p. 8.

Q 221. Rhombic Dodecahedrons

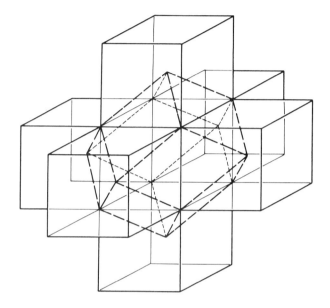

In a compact aggregation of equal cubes (which are space filling) pass planes through the six pairs of opposite edges of alternate cubes. Thus these cubes are dissected into six congruent pyramids with square bases and lateral edges equal to one-half the space diagonal of the cube. Each of the nondissected cubes is now faced with six pyramids which together with the cube constitute a rhombic dodecahedron (one face for each edge of the cube) with the edges of the cube as face diagonals. Since the volumes of all the cubes are used in this new assemblage, the rhombic dodecahedrons are space filling.

It follows immediately that the volume of a rhombic dodecahedron is equal to twice the cube of the short diagonal of a face.

Q 222. A Fibonacci Relationship

Repeatedly applying the recursion relationship, we have

$$F_n = F_{n-1} + F_{n-2} = F_{n-2} + F_{n-3} + F_{n-3} + F_{n-4}$$

$$= F_{n-3} + F_{n-4} + 2F_{n-4} + 2F_{n-5} + F_{n-4}$$

$$= 5F_{n-4} + 3F_{n-5}.$$

Now $F_5 = 5$, so every fifth number is divisible by 5.

—E. M. Scheuer, *A.M.M.*, 67 (August, 1960), 694.

That is 1, 1, 2, 3, $\underline{5}$, 8, 13, 21, 34, $\underline{55}$, 89, 144, 233, 377, $\underline{610}$, $\cdots$

Q 223. Cryptic Multiplication

Since $(188)(8) = 1504$, the first O must be greater than 1.   That particular O times the first E of the multiplier ≤ 8, so that $O = 3$ and that $E = 2$.   The only values of $3EE$ which produce an EOE pattern upon multiplication by 2 are 306, 308, 326, 328, 346, and 348. None of these when multiplied by 4 or 6, and only the last two when multiplied by 8, produce an $EOEE$ pattern.   Now $(346)(28) = 9688$, so the unique solution is

$$
\begin{array}{r}
348 \\
28 \\
\hline
2784 \\
696 \\
\hline
9744
\end{array}
$$

—W. Fitch Cheney.

Q 224. Creased Rectangle

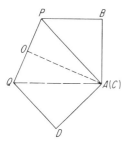 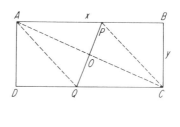

Let the rectangle be $ABCD$, with length $x = AB = DC$, and width $y = AD = BC$, where $x \geqq y$. Bring C into coincidence with A and let the crease be PQ. By symmetry, $PC = PA = CQ = AQ$, so AC perpendicularly bisects PQ at O. Now triangles AOP and ABC are similar, so

$$PO/BC = AO/AB$$

and

$$PQ = 2PO = 2(AO)(BC)/AB = (y/x)\sqrt{x^2 + y^2}.$$

—Alan Wayne, *S.S.M.*, 64 (March, 1964), 241.

Q 225. A Man's Birthdate

A man living in 1937 could not have been 43 years old in 1849, that is, $(43)^2$. Therefore, he must have been 44 years old in 1936. From the conditions given,

$$44 + m = d^2, \qquad 0 < m < 13.$$

The only integral solution is $m = 5$, $d = 7$. The man was born May 7, 1892.

—Lucille G. Meyer, *N.M.M.*, 11 (March, 1937), 282.

183

Q 226. A Fractional Equation

It is evident upon inspection that 0 and $a + b$ are roots. Furthermore, in general, if $m + n = 1/m + 1/n$, then $(m + n)(mn - 1) = 0$. Consequently, $(x - a)/b + (x - b)/a = 0$, or $(a + b)x = a^2 + b^2$, so the third root of the cubic is $(a^2 + b^2)/(a + b)$.

Q 227. A Particular Isosceles Triangle

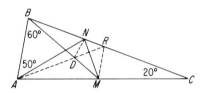

Since the sum of the angles of a triangle equals a straight angle or $180°$, then $\angle CBA = \angle CAB = 80°$, $\angle CBM = 20°$, and $\angle BAN = 50° = \angle BNA$ (so $BN = AB$). Draw MR parallel to AB, and AR intersecting BM at D. Draw ND. By symmetry, triangles ABD and DRM are isosceles and hence equiangular. Then $BD = AB = BN$, so $\angle BND = \angle BDN = 80°$, and $\angle NDR = 40°$. Now $\angle MRC = 80°$, so $\angle NRD = 40° = \angle NDR$, and $ND = NR$. Then since $DM = MR$, NM is the perpendicular bisector of DR. Therefore $\angle BMN = 60°/2$ or $30°$.

—*M.M.*, 39 (September, 1966), 253.

Q 228. Tree Leaves

Since each of the n distinct nonnegative integers, $0, 1, 2, \cdots,$ $(n - 1)$ is less than n, the statement is obviously incorrect. However, if it be required also that no tree be completely devoid of leaves, then the statement would be true. Since each of the set of $n - 1$ distinct integers, 1, 2, $\cdots$, $(n - 1)$ is less than n, an nth positive integer less than n must duplicate one of the set.

Q 229. A Diophantine Cubic

One solution is obviously $(x, y) = (-1, 0)$. Furthermore, x cannot have any other negative value, since y^2 is positive. We may write

$$x^3 + 1 = (x + 1)(x^2 - x + 1) = (x + 1)^2\left(x - 2 + \frac{3}{x + 1}\right) = y^2.$$

Now 0 and 2 are the only non-negative integers for which $3/(x + 1)$ is an integer. Consequently, the only solutions other than $(-1,\ 0)$ are $(0,\ \pm 1)$ and $(2,\ \pm 3)$.

Q 230. Countries on a Sphere

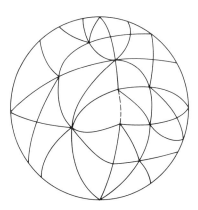

Suppose that the graph G is such a triangulation. If the boundary

185

line joining the two odd vertices is removed (forming a four-sided country), then every vertex in the resulting graph G' is even. Hence we can color the countries of G' with two colors, say red and black, so that adjoining countries have different colors. Let r and b denote the number of countries of G' with colors red and black, respectively. We may suppose that the one country with four sides is colored red. Since all the remaining countries have three sides it follows that $b = [4 + (r - 1)3]/3$. But this is not an integer, so no such triangulation exists.

—J. W. Moon, *A.M.M.*, 72 (January, 1965), 81.

Q 231. Henry's Trip

The trip took six hours. Suppose the hour hand had a backward extension. If the minute hand and hour hand were coincident at the start, after *six hours* the regular hand and the extension would change places and the minute hand, after exactly six circuits, would return to its starting position to coincide with the extension.

—Charles Salkind, *M.M.*, 28 (March, 1955), 241.

Q 232. A Power Series

We have

$$1/(1 + x)(1 + x^2)(1 + x^4)(1 + x^8)$$
$$= (1 - x)/(1 - x^{16})$$
$$= (1 - x)(1 + x^{16} + x^{32} + x^{48} + \cdots)$$
$$= 1 - x + x^{16} - x^{17} + x^{32} - x^{33} + \cdots.$$

—M. S. Klamkin, *M.M.*, 29 (September, 1955), 53.

Q 233. Parallels in a Triangle

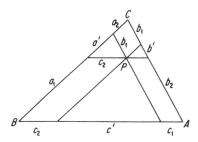

From the parallelograms and similar triangles in the figure, we have $a'/a = b_1/b$, $a'/a = c_2/c$, $b'/b = c_1/c$, $b'/b = a_2/a$, $c'/c = b_2/b$, $c'/c = a_1/a$. These together with the identities $a'/a = a'/a$, $b'/b = b'/b$, $c'/c = c'/c$ constitute nine equations which when added together give

$$3(a'/a + b'/b + c'/c)$$

$$= (a_1 + a' + a_2)/a + (b_1 + b' + b_2)/b + (c_1 + c' + c_2)/c = 3.$$

Therefore, $a'/a + b'/b + c'/c = 1$.

—*S.S.M.*, 55 (November, 1955), 660.

Q 234. A Partitioning Problem

We have $316/_{11} = 28$ with a remainder of 8, and $8/(13 - 11) = 4$. Hence, the two parts are $(4)(13)$ or 52 and 264.

316 can also be partitioned into $52 + (11)(13)$ and $264 - (11)(13)$ or 195 and 121.

Q 235. Fair Fares

In the scale of six,

$1^2 = 1$, $2^2 = 4$, $3^2 = 13$, $4^2 = 24$, and $5^2 = 41$.   Hence $F = 1$.

Also, $11^2 = 121$, $22^2 = 524$, $33^2 = 2013$, $44^2 = 3344$, and $55^2 = 5401$, so $E = 2$.   Therefore $(122)^2 = 15,324$.

An interesting by-product is $(221)^2 = 53,241$.

Q 236. How Old Is Willie?

Willie's friend, dispensing with trial-and-error, might make use of the fact that $a - b$ is an integral divisor of $P(a) - P(b)$ when a, b are distinct integers and $P(x)$ is a polynomial with integral coefficients.   Denoting the "larger integer" tried by N and Willie's age by A, we have $N - 7$ divides $(85 - 77)$ or 8, $A - 7$ divides 77, $A - N$ divides 85 and $7 < N < A$.   It follows that N must be one of 8, 9, 11, 15 and A one of 14, 18, 84.   Since $A - N$ divides 85, the second integer tried must have been 9 and Willie is fourteen years old.

—D. C. B. Marsh, *A.M.M.*, 64 (October, 1957), 593.

The polynomial must have been of the form

$$(x - 7)(x - 9)(x - 14)Q(x) - 3x^2 + 52x - 140.$$

Q 237. Accelerating Particle

If we plot v against t, the area under the curve must be the same (1 square unit) as that of an isosceles triangle having the same base

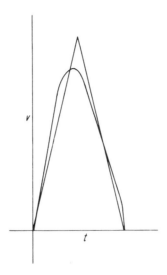

and an altitude of 2. The slopes of the sides of the triangle are ± 4. Part of the v, t curve must fall outside the triangle or coincide with its sides. Thus the slope, a, of the curve is numerically $\geqq 4$ at some point.

—P.M.E.J., 1 (November, 1952), 280.

Q 238. Stamps for Buck

The 1- and 2-cent stamps were purchased in 12-cent lots, and the amount spent for them had to be a multiple of 5, namely, 60 cents. Hence, five 2-cent, fifty 1-cent, and eight 5-cent stamps were purchased.

—M.M., 32 (January, 1959), 171.

Q 239. A Twenty Question Game

If a questioner wishes to determine an object previously chosen from a finite set of objects, his most efficient procedure is to ask at each stage whether the chosen object has a property which is possessed by exactly half of the members of the set. Regardless of which answer he receives, the set containing the chosen object is thereby cut in half. By this procedure the questioner can determine a chosen number in twenty questions, provided that the number chosen is a positive integer not greater than 2^{20}.

The ith question for $i = 1, 2, \cdots, 20$, could be: "If the number is written to the base two, is the ith digit the number 1?" Notice that if the answer is always "no" then the number (base two) is the 21-digit number consisting of 1 followed by 20 zeros. If at least one answer is "yes" then the number contains at most twenty digits, each of which has been determined.

—H. M. Gehman, *A.M.M.*, 58 (January, 1951), 40.

Q 240. Inscribed Spheres

The hexahedron is composed of two regular tetrahedrons, with volumes V_t. Now $V_t/V_o = \frac{1}{4}$. [See Q 113.] Hence, $V_h/V_o = \frac{1}{2}$. Also, the surface of the hexahedron is $\frac{6}{8}$ that of the octahedron, or $S_h/S_o = \frac{3}{4}$. Since $V_h = r_h S_h/3$ and $V_o = r_o S_o/3$, it follows that $r_h/r_o = (V_h/V_o)(S_o/S_h) = \frac{2}{3}$.

—Howard Eves, *A.M.M.*, 56 (December, 1949), 693.

Q 241. Disks From a Disk

Stewart's theorem states that the square of the distance of a point on the base of a triangle from the opposite vertex multiplied by the

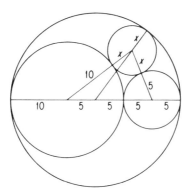

base is equal to the sum of the squares of the other two sides, each multiplied by the nonadjacent segment of the base, less the product of these two segments multiplied by the base.

When this theorem is applied to the triangle formed by the lines of centers in the figure, we have

$$15(15 - x)^2 = 10(10 + x)^2 + 5(5 + x)^2 - (15)(10)(5).$$

This equation simplifies to $700x = 3000$. Hence the largest disk that can be cut from the remainder of the plywood is one with a radius of $3\tfrac{0}{7}$ inches. The width of the saw blade has been neglected throughout the problem.

—*S.S.M.*, 59 (April, 1959), 326.

Q 242. Sum of Squares of Binomial Coefficients

The possible ways in which n objects can be chosen from n red and n black objects is $C(2n, n)$ or $[C(n, 0)][C(n, n)] + [C(n, 1)] \times [C(n, n - 1)] + [C(n, 2)][C(n, n - 2)] + \cdots + [C(n, n - 1)] \times$

$[C(n, 1)] + [C(n, n)][C(n, 0)]$.   But $C(n, k) = C(n, n - k)$, so we have

$$C(2n, n) = \sum_{i=0}^{n} [C(n, i)]^2.$$

—*M.M.*, 24 (September, 1950), 54.

This is equivalent to stating that the sum of the squares of the integers in the kth cross diagonal of the Pascal triangle (see 179) is equal to the kth (or middle) number in the $(2k - 1)$st cross diagonal. For another quick proof of this, see Harry Siller, *A.M.M.*, 42 (January, 1935), 46.

Q 243. Christmas Cryptarithm

With the aid of a table of squares it is readily determined that *ALL* is either 100, 144, 400, or 900, and *TO* is either 36 or 81.

The only four-digit square with a square digit sum and 1, 4, or 9 as a tens' digit is 7396 = *XMAS*, so *ALL* = 900.   To avoid double representation of digits *TO* must be 81.   Then *MERRY* is either 35224 or 34225, but only the latter is a square number.   Thus the numerical interpretation of the greeting is

9   34225   7396   81   900

If the restriction on the sum of the digits of the squares is removed, there is a second solution,

4   27556   3249   81   400.

Q 244. Centers of Gravity

(a) Clearly the center of gravity of a semicircular wire lies on the radius perpendicular to the diameter at a distance y from the diameter. Rotating the wire about the diameter generates a spherical surface. Pappus' first theorem states that the area of a surface of revolution, formed by revolving a curve about a line in its plane not cutting the curve, is equal to the product of the length of the generating curve and the circumference of the circle described by the centroid of the curve. Thus, $4\pi r^2 = (\pi r)(2\pi y)$, so $y = 2r/\pi$.

—M. S. Klamkin, *M.M.*, 26 (March, 1953), 226.

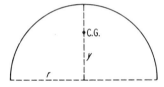 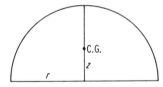

(b) Pappus' second theorem states that the volume of a solid of revolution, formed by revolving a plane area about a line in its plane not cutting the area, is equal to the product of the generating area and the circumference of the circle described by the centroid of the area.

Clearly the center of gravity of the semicircular area falls on the radius perpendicular to the diameter at a distance z from the diameter. Rotating the area about the diameter generates a spherical volume. Thus, $4\pi r^3/3 = (\pi r^2/2)(2\pi z)$, so $z = 4r/3\pi$.

—M. S. Klamkin, *M.M.*, 27 (March, 1954), 227.

Q 245. Two Equal Triads

If $x + y + z = a + b + c$ and $xyz = abc$, then

$$abc(ab + bc + ca - xy - yz - zx) = bc(x - a)(y - a)(z - a)$$

$$= ca(x - b)(y - b)(z - b) = ab(x - c)(y - c)(z - c).$$

If no factor in any of the last three expressions is zero, then one expression has all positive factors and another has exactly three negative factors. Then the equalities could not hold. Thus each expression has one zero factor, and in some order x, y, z are equal to a, b, c.

—W. J. Blundon, *A.M.M.*, 72 (February, 1965), 185.

Q 246. An Irrational Sum

The exponents form the sequence $-1, -4, -8, -13, -19, -26,$ $\cdots$ with the difference of consecutive terms constantly increasing by 1. Written to the base six, the sum of the series is the nonrepeating decimal, $0.1001000100001000001\cdots$, which is irrational.

—David L. Silverman, *M.M.*, 32 (March, 1959), 229.

Q 247. Examination of Six Students

Given n students, the probability that the first man finished will not have to pass over any of the others in gaining the aisle is obviously $2/n$. Hence, the required probability that one of the six students will have to pass over one or more students is $1 - (\frac{2}{6})(\frac{2}{5})(\frac{2}{4})(\frac{2}{3})$ or $\frac{43}{45}$.

—Howard Eves, *A.M.M.*, 50 (March, 1943), 202.

Q 248. Construction by Compasses Alone

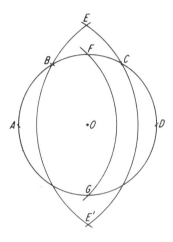

With radius r and center O, draw a circle.   With the same radius strike off equal arcs AB, BC, CD.   Then AD is a diameter.   With radius AC and A and D as centers draw circles intersecting in E and E'.   With radius OE and A as center draw a circle intersecting the original circle at F and G.   Then $AFDG$ is an inscribed square.

—John A. Dyer, *P.M.E.J.*, 1 (April, 1950), 55.

Proof: $AC = r\sqrt{3} = AE.$ $OE = \sqrt{(AE)^2 - r^2} = r\sqrt{2} = AF.$

Q 249. The Bonus Fund

In effect, five dollars was taken from $95/5$ or 19 persons, so the fund contained $20(50) - 5$ or 995 dollars.

Q 250. The Court Mathematician's Salary

As pointed out by L. S. Frierson in Chapter 5 of W. S. Andrews' *Magic Squares and Cubes,* the elements of any third-order magic square may be represented with three parameters, substantially as shown on the left below. It is clear that these nine elements can be rearranged into another array in which the rows are arithmetic progressions with the same common difference, and also the columns are arithmetic progressions having the same d. The central element is the mean of four different pairs of elements as well as of the square.

$(e+x)$	$(e-x-y)$	$(e+y)$		$(e-x-y)$	$(e-y)$	$(e+x-y)$
$(e-x+y)$	e	$(e+x-y)$		$(e-x)$	e	$(e+x)$
$(e-y)$	$(e+x+y)$	$(e-x)$		$(e-x+y)$	$(e+y)$	$(e+x+y)$

It is desired to form a magic square of order three from the set of primes p_i having the property that the $p_i + 2$ are also prime. Now unless 3 or 5 appears as an element, the p_i must terminate with 9, 7, or 1. Then e must be the prime mean of two primes terminating with 9, 9 or 7, 1; or 7, 7; or 1, 1, respectively. Consequently we examine the sequence formed by the smaller member of the successive twin primes. That is:

3, 5, 11, 17, 29, 41, 59, 71, 101, 107, 137, 149, 179, 191, 197, 227, 239, 269, 281, $\cdots$.

The smallest value of e ending in 9 and which is the mean of at least four prime pairs is determined by

$$2(149) = 107 + 191 = 101 + 197 = 71 + 227 = 59 + 239$$
$$= 29 + 269 = 17 + 281.$$

(For any prime < 149 and ending in 1 or 7 there are not more than three smaller primes in the sequence with the same terminal digit.)

When we successively fit these pairs into the middle row of the arithmetic progression array, we find that there is only one arrangement that permits three of the other pairs to be placed properly, namely:

$$\begin{array}{ccc} 17 & 59 & 101 \\ 107 & 149 & 191 \\ 197 & 239 & 281 \end{array}$$

This leads immediately to the three magic squares involved:

$$\begin{array}{ccc} 191 & 17 & 239 \\ 197 & 149 & 101 \\ 59 & 281 & 107 \end{array} \qquad \begin{array}{ccc} 192 & 18 & 240 \\ 198 & 150 & 102 \\ 60 & 282 & 108 \end{array} \qquad \begin{array}{ccc} 193 & 19 & 241 \\ 199 & 151 & 103 \\ 61 & 283 & 109 \end{array}$$

The mathematician's salary was $9e + 9 = 9(149 + 1)$ or 1350 "dollars."

This result was arrived at by another route in *A.M.M.*, 55 (September, 1948), 429. The next solution, by Wm. H. Benson, is 2088, 1392, 4800; 5442, 2730, 18; 660, 4158, 3372; total, 24,570 "dollars."

Q 251. Packing Cylinders

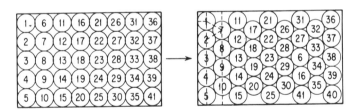

Two cylinders must be removed. Number the cylinders as shown in the figure. Remove cylinders 6 and 16.

Move 7–10 up and left to the new positions shown. Move 11–15 to left. Move 17–20 up and left. Move 21–25 to left. The remaining cylinders may be maneuvered in a variety of ways. For example: Move 26, 28, 29, 30, 31 to the left; push 32 against 31 and 37; 27 against 28 and 32; 26 against 21 and 22.   Move 31 to the left, 32 against 36 and 37; 27 up against 31; 28, 29, 30 up and to the left; 33, 34, 35 up and to the right.   Insert 6, 16, and 41 into the positions shown.

When the cylinders are snugged against each other in alternate columns of five and four, the lines of centers of the columns are $\sqrt{3}/2$ apart.   The overall length of the pack is $8(\sqrt{3}/2) + 1 = 4\sqrt{3} + 1 \doteq 7.928 < 8$, so the forty-one cylinders will rattle in the box.

Q 252. Leg of a Pythagorean Triangle

Fermat's theorem states that if p is a prime and m is not divisible by p, then $m^{p-1} \equiv 1 \pmod{p}$.   The relationship $a^2 + b^2 = c^2$ may be written as

$$(a^{3-1} - 1) + (b^{3-1} - 1) = c^2 - 2 \quad \text{or} \quad (c^{3-1} - 1) - 1.$$

If neither a nor b is divisible by 3, then each of the terms on the left-hand side of the equation is divisible by 3.   But, whether c is divisible by 3 or not, the right-hand side of the equation leaves a remainder upon division by 3.   Hence, for the equation to hold, one of a, b must be a multiple of 3.

Q 253. Powers of Two

In the binary system this sum is represented by n 1's followed by a zero.   Addition of two 1's converts the sum into a 1 followed by $n + 1$ zeros.   Hence the sum of the series is $2^{n+1} - 2$.

Q 254. Colorful Square Arrays

There must be at least three distinct patterns, each with a different color for the central square.

There are three ways to select the color of the cell in the upper left-hand corner. For each of these there are two ways to color the upper center cell, and then there are two ways to color the left-hand cell of the middle row. Once these cells have been colored, there is only one possible choice for each of the remaining cells. Thus there are (3)(2)(2) or 12 ways to color the fixed array. Any one of these arrangements may be rotated to coincide with three others, so there are only 12/4 or 3 distinct basic patterns.

—Wm. H. Benson.

Q 255. Bisected Parallelogram

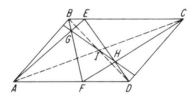

Pappus' theorem states that if the vertices of a hexagon fall alternately on two lines, the intersections of opposite sides are collinear. If AB and DC intersect at I, then the opposite sides of the hexagon $AEDBFCA$ intersect at G, I, and H, which are collinear. Any line through I, the intersection of the diagonals, will bisect the parallelogram.

—Mannis Charosh, *M.M.*, 38 (September, 1965), 252.

Q 256. Squares of Reverse Integers

$12345 \leq N^2 \leq 54321$, so $113 \leq N \leq 221$.   The reversal of N also must fall within this range. $N^2 \equiv 0 \pmod 5$, so $N \equiv 0 \pmod 5$. The only three-digit integers congruent to zero within the restricted range are: 113, 122, 131, 140, 145, 154, 203, 212, and 221.   There is only one pair of reverse integers in this set. Hence the unique solution is $(221)^2 = 53241$ which leads to $15324 = (122)^2$.

As a serendipity we find $(203)^2 = 42013$, another permutation of consecutive digits.

Q 257. A Questionable Sum

Without loss of generality, a and b may be taken as relatively prime integers, $a > b$.   If $a/b + b/a = k$, an integer, then $a^2 + b^2 = abk$.   That is, $b^2 = a(bk - a)$, so a divides b^2, an impossibility except when $a = \pm b$.

Q 258. Is the Square Fault-Free?

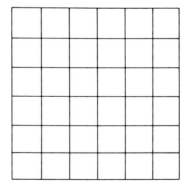

No fault-free arrangement is possible.   Consider a 6-inch by 6-inch grid, which has five vertical and five horizontal grid lines.   When covered by the eighteen 2-inch by 1-inch dominoes, the assemblage will be fault-free if each grid line intersects at least one domino.

Every vertical grid line has an even number of squares to its left. Since each uncut domino occupies an even number of squares, parts of any cut dominoes also must occupy an even number of squares to the left of the line.   Thus, if it is not to be a fault line, every grid line must cut at least two dominoes.   No domino can be cut by more than one grid line, so a total of 20 dominoes need to be cut by the ten grid lines.   But only 18 dominoes are available, so at least one grid line must be a fault line.

—Solomon W. Golomb, *Scientific American* (December, 1960), p. 168.

Q 259. Relatively Prime Numbers

Represent the greatest common divisor of x and y by (x, y).   Then

$$(35, 58) = (35, 23) = (12, 23) = (12, 11) = (1, 11) = 1.$$

Hence 35 and 58 are relatively prime in any system of notation having a base > 8.

—David L. Silverman, *M.M.*, 38 (November, 1965), 326.

Q 260. Five Consecutive Integers

Fourth powers of even integers are of the form $4k$, and those of odd integers are of the form $4k + 1$.   Hence the sum of four consecutive fourth powers is of the form $4k + 2$, which cannot be a fourth power.

Q 261. Polygonal Path in a Lattice

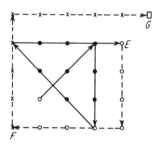

The figure shows that for $N = 3$ a path of $2N - 2$ or 4 segments (through the dots) can be drawn ending at E. Two additional segments along the dotted lines ending in F suffice for $N = 4$, and two more segments ending in G suffice for $N = 5$. Indeed, the process of adding 2 segments for each unit increase in N may be continued indefinitely.

—M. S. Klamkin, *A.M.M.*, 62 (February, 1955), 124.

Q 262. Wanted—Integer Solutions

Multiply the given equation by 4 and add 1 to each side to get

$$4x^4 + 4x^3 + 4x^2 + 4x + 1 = (2y + 1)^2.$$

For $x = -1$, $y = -1$ or 0; for $x = 0$, $y = -1$ or 0; for $x = 2$, $y = -6$ or 5; for $x = 1$, y is nonintegral.

These are the only six integral solutions of the equation, since for all $x < -1$ or $x > 2$, the left-hand side of the equation is greater than $(2x^2 + x)^2$ but less than $(2x^2 + x + 1)^2$, so cannot be an integral square for integral x as it would have to be to equal the right-hand side.

—D. C. B. Marsh, A.M.M., 73 (October, 1966), 895.

Q 263. A Composite Number

$q = (p_1 + p_2)/2$ is the arithmetic mean of p_1 and p_2, so $p_1 < q < p_2$. But p_1 and p_2 are consecutive primes, so q must be composite.

—John D. Baum, M.M., 39 (May, 1966), 196.

Q 264. Four Simultaneous Linear Equations

Interchange of x and u and of y and v in (1) and (2) produces (4) and (3), respectively, except that the sign of the right-hand member is changed.   Consequently, $u = -x$ and $v = -y$.   When the indicated substitutions are made in (1) and (2) we have

$$
\begin{aligned}
-4x + 4y &= 16 \\
6x - 2y &= -16 \\
\hline
8x &= -16
\end{aligned}
$$

So, $x = -2, y = 2, v = -2, u = 2$.

Q 265. Property of a Quadrangle

Triangles having the same altitudes are to each other as their

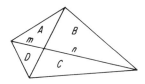

bases, so if the four triangles are named in cyclic order, we have

$$\frac{A}{A+B} = \frac{m}{m+n} = \frac{D}{C+D} = \frac{A+D}{A+B+C+D} = \frac{D+A}{Q}.$$

In like manner, $B/(B+C) = (A+B)/Q$, $C/(C+D) = (B+C)/Q$, and $D/(D+A) = (C+D)/Q$.

Multiplication of the four equalities gives

$$(A)(B)(C)(D) = (A+B)^2(B+C)^2(C+D)^2(D+A)^2/Q^4.$$

—Leon Bankoff, M.M., 38 (September, 1965), 248.

Q 266. When Is the Division Exact?

$$f(n) = \frac{n^4 + n^2}{2n+1} = \frac{n^2(n^2+1)}{2n+1}$$

$$= \frac{n^2}{4}\left[\frac{4n^2+4}{2n+1}\right] = \left(\frac{n}{2}\right)^2\left[2n - 1 + \frac{5}{2n+1}\right].$$

Clearly n and $2n + 1$ have no common factor other than 1. So $f(n)$ cannot be an integer unless $5/(2n + 1)$ is, that is, when $n = 2$, 0, -1, or -3. For the only *positive* integral value, $n = 2$, $f(n) = 4$.

Q 267. A Doubtful Equation

For the statement to be true as it stands, the numbers must be expressed in different scales of notation, that is, $342_a = 97_b$. If $b = 10$, then since $3(4)^2 = 48$ and $3(6)^2 = 108$, $a = 5$. Indeed, $3(5)^2 + 4(5) + 2 = 97$.

In general, if $3a^2 + 4a + 2 = 9b + 7$, then $b = (3a^2 + 4a - 5)/9$ which is an integer only if a has the form $9x + 5$, whereupon

$b = 27x^2 + 34x + 10$. Thus there are any number of solutions, for example: $342_5 = 97_{10}$, $342_{14} = 97_{71}$, and so on.

Q 268. Dissected Dodecagon

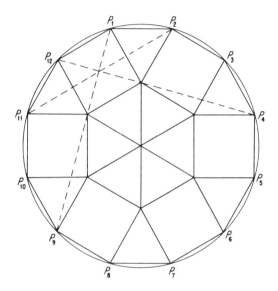

(a) By considering the angles formed as angles inscribed in the circumscribed circle, it is evident that the nine diagonals from any vertex divide the 150° angle at that vertex into ten equal 15° angles.

Draw portions of the diagonals P_1P_6, P_2P_9, P_3P_8, P_4P_{11}, P_5P_{10}, P_7P_{12} until they intersect the diagonals from the adjacent vertices. Angles $P_4P_3P_8$, $P_3P_4P_{11}$, etc., are equal to 60°, so equilateral triangles have been constructed internally on alternate sides of the dodecagon. It follows that their vertices are also vertices of squares on the other

205

sides, since the angles $P_2P_3P_8$, $P_3P_2P_9$, etc., are $90°$ angles. The fourth sides of the squares are sides of a regular hexagon, whose diagonals divide it into equilateral triangles. Thus the dodecagon is dissected into 12 congruent equilateral triangles and 6 congruent squares.

(*b*) Angles $P_1P_{12}P_4$ and $P_{12}P_1P_9$ are $45°$ each, so P_1P_9 and P_4P_{12} are diagonals of a square. P_2P_{11} is an axis of symmetry in the equilateral convex hexagon formed by two equilateral triangles and a square, so it passes through the center of the square and is concurrent with P_1P_9 and P_4P_{12}.

Q 269. No Real Roots

If there is a real root it must be negative, say $-y$. But

$$1 - y + y^2/2! - y^3/3! + \cdots + y^{2n}/(2n)! > e^{-y} > 0.$$

Hence, the equation has no real roots.

—Joe Lipman, *A.M.M.*, 67 (April, 1960), 379.

Q 270. Impossible Cube

Since $ar^2 + ar + a = a^3$, then $r^2 + r + 1 = a^2$. Now any digit, a, in a system of numeration is less than the base, r. Hence the situation is impossible.

—Charles McCracken, Jr., *S.S.M.*, 52 (March, 1952), 241.

NAMES MENTIONED

Numerical references are to Challenge Problems (such as 131) or to Quickie Solutions (such as Q 51) and not to pages.

Adler, Claire	Q 35
Alfred, Brother U.	Q 131
Altshiller-Court, Nathan	Q 82
Andrews, W. S.	Q 250
Anning, Norman	Q 30, Q 63, Q 92, Q 168, Q 172, Q 199
Anselone, P. M.	Q 205
Arena, J. F.	Q 129
Aude, H. T. R.	Q 139
Ball, W. W. R.	220
Bankoff, Leon	Q 82, Q 116, Q 265
Baum, J. D.	Q 263
Beberman, Max	Q 45
Benson, Wm. H.	Q 156, Q 250, Q 254
Blundon, W. J.	Q 245
Brown, Bancroft H.	Q 167
Buchman, Aaron	Q 72, Q 103, Q 138, Q 192
Buckley, D. G.	Q 187
Buker, W. E.	Q 123
Bush, L. E.	Q 104, Q 137
Butchart, J. H.	Q 15
Carver, W. B.	Q 31
Ceva, G.	Q 89
Charosh, Mannis	Q 255
Chase, Louis R.	Q 198, Q 204
Cheney, W. Fitch	Q 128, Q 223

A CATALOG OF SELECTED DOVER
BOOKS IN ALL FIELDS OF INTEREST

CONCERNING THE SPIRITUAL IN ART, Wassily Kandinsky. Pioneering work by father of abstract art. Thoughts on color theory, nature of art. Analysis of earlier masters. 12 illustrations. 80pp. of text. 5⅜ x 8½.   23411-8

ANIMALS: 1,419 Copyright-Free Illustrations of Mammals, Birds, Fish, Insects, etc., Jim Harter (ed.). Clear wood engravings present, in extremely lifelike poses, over 1,000 species of animals. One of the most extensive pictorial sourcebooks of its kind. Captions. Index. 284pp. 9 x 12.   23766-4

CELTIC ART: The Methods of Construction, George Bain. Simple geometric techniques for making Celtic interlacements, spirals, Kells-type initials, animals, humans, etc. Over 500 illustrations. 160pp. 9 x 12. (Available in U.S. only.)   22923-8

AN ATLAS OF ANATOMY FOR ARTISTS, Fritz Schider. Most thorough reference work on art anatomy in the world. Hundreds of illustrations, including selections from works by Vesalius, Leonardo, Goya, Ingres, Michelangelo, others. 593 illustrations. 192pp. 7⅛ x 10¼.   20241-0

CELTIC HAND STROKE-BY-STROKE (Irish Half-Uncial from "The Book of Kells"): An Arthur Baker Calligraphy Manual, Arthur Baker. Complete guide to creating each letter of the alphabet in distinctive Celtic manner. Covers hand position, strokes, pens, inks, paper, more. Illustrated. 48pp. 8¼ x 11.   24336-2

EASY ORIGAMI, John Montroll. Charming collection of 32 projects (hat, cup, pelican, piano, swan, many more) specially designed for the novice origami hobbyist. Clearly illustrated easy-to-follow instructions insure that even beginning papercrafters will achieve successful results. 48pp. 8¼ x 11.   27298-2

THE COMPLETE BOOK OF BIRDHOUSE CONSTRUCTION FOR WOODWORKERS, Scott D. Campbell. Detailed instructions, illustrations, tables. Also data on bird habitat and instinct patterns. Bibliography. 3 tables. 63 illustrations in 15 figures. 48pp. 5¼ x 8½.   24407-5

BLOOMINGDALE'S ILLUSTRATED 1886 CATALOG: Fashions, Dry Goods and Housewares, Bloomingdale Brothers. Famed merchants' extremely rare catalog depicting about 1,700 products: clothing, housewares, firearms, dry goods, jewelry, more. Invaluable for dating, identifying vintage items. Also, copyright-free graphics for artists, designers. Co-published with Henry Ford Museum & Greenfield Village. 160pp. 8¼ x 11.   25780-0

HISTORIC COSTUME IN PICTURES, Braun & Schneider. Over 1,450 costumed figures in clearly detailed engravings–from dawn of civilization to end of 19th century. Captions. Many folk costumes. 256pp. 8⅜ x 11¾.   23150-X

STICKLEY CRAFTSMAN FURNITURE CATALOGS, Gustav Stickley and L. & J. G. Stickley. Beautiful, functional furniture in two authentic catalogs from 1910. 594 illustrations, including 277 photos, show settles, rockers, armchairs, reclining chairs, bookcases, desks, tables. 183pp. 6½ x 9¼.   23838-5

AMERICAN LOCOMOTIVES IN HISTORIC PHOTOGRAPHS: 1858 to 1949, Ron Ziel (ed.). A rare collection of 126 meticulously detailed official photographs, called "builder portraits," of American locomotives that majestically chronicle the rise of steam locomotive power in America. Introduction. Detailed captions. xi+ 129pp. 9 x 12.   27393-8

AMERICA'S LIGHTHOUSES: An Illustrated History, Francis Ross Holland, Jr. Delightfully written, profusely illustrated fact-filled survey of over 200 American light-houses since 1716. History, anecdotes, technological advances, more. 240pp. 8 x 10¾.
                    25576-X

TOWARDS A NEW ARCHITECTURE, Le Corbusier. Pioneering manifesto by founder of "International School." Technical and aesthetic theories, views of industry, economics, relation of form to function, "mass-production split" and much more. Profusely illustrated. 320pp. 6⅛ x 9¼. (Available in U.S. only.)   25023-7

HOW THE OTHER HALF LIVES, Jacob Riis. Famous journalistic record, exposing poverty and degradation of New York slums around 1900, by major social reformer. 100 striking and influential photographs. 233pp. 10 x 7⅞.   22012-5

FRUIT KEY AND TWIG KEY TO TREES AND SHRUBS, William M. Harlow. One of the handiest and most widely used identification aids. Fruit key covers 120 deciduous and evergreen species; twig key 160 deciduous species. Easily used. Over 300 photographs. 126pp. 5⅜ x 8½.   20511-8

COMMON BIRD SONGS, Dr. Donald J. Borror. Songs of 60 most common U.S. birds: robins, sparrows, cardinals, bluejays, finches, more–arranged in order of increasing complexity. Up to 9 variations of songs of each species.
                    Cassette and manual 99911-4

ORCHIDS AS HOUSE PLANTS, Rebecca Tyson Northen. Grow cattleyas and many other kinds of orchids–in a window, in a case, or under artificial light. 63 illustrations. 148pp. 5⅜ x 8½.   23261-1

MONSTER MAZES, Dave Phillips. Masterful mazes at four levels of difficulty. Avoid deadly perils and evil creatures to find magical treasures. Solutions for all 32 exciting illustrated puzzles. 48pp. 8¼ x 11.   26005-4

MOZART'S DON GIOVANNI (DOVER OPERA LIBRETTO SERIES), Wolfgang Amadeus Mozart. Introduced and translated by Ellen H. Bleiler. Standard Italian libretto, with complete English translation. Convenient and thoroughly portable–an ideal companion for reading along with a recording or the performance itself. Introduction. List of characters. Plot summary. 121pp. 5¼ x 8½.   24944-1

TECHNICAL MANUAL AND DICTIONARY OF CLASSICAL BALLET, Gail Grant. Defines, explains, comments on steps, movements, poses and concepts. 15-page pictorial section. Basic book for student, viewer. 127pp. 5⅜ x 8½.   21843-0

THE CLARINET AND CLARINET PLAYING, David Pino. Lively, comprehensive work features suggestions about technique, musicianship, and musical interpretation, as well as guidelines for teaching, making your own reeds, and preparing for public performance. Includes an intriguing look at clarinet history. "A godsend," *The Clarinet,* Journal of the International Clarinet Society. Appendixes. 7 illus. 320pp. 5⅜ x 8½. 40270-3

HOLLYWOOD GLAMOR PORTRAITS, John Kobal (ed.). 145 photos from 1926-49. Harlow, Gable, Bogart, Bacall; 94 stars in all. Full background on photographers, technical aspects. 160pp. 8⅜ x 11¼. 23352-9

THE ANNOTATED CASEY AT THE BAT: A Collection of Ballads about the Mighty Casey/Third, Revised Edition, Martin Gardner (ed.). Amusing sequels and parodies of one of America's best-loved poems: Casey's Revenge, Why Casey Whiffed, Casey's Sister at the Bat, others. 256pp. 5⅜ x 8½. 28598-7

THE RAVEN AND OTHER FAVORITE POEMS, Edgar Allan Poe. Over 40 of the author's most memorable poems: "The Bells," "Ulalume," "Israfel," "To Helen," "The Conqueror Worm," "Eldorado," "Annabel Lee," many more. Alphabetic lists of titles and first lines. 64pp. 5$\frac{3}{16}$ x 8¼. 26685-0

PERSONAL MEMOIRS OF U. S. GRANT, Ulysses Simpson Grant. Intelligent, deeply moving firsthand account of Civil War campaigns, considered by many the finest military memoirs ever written. Includes letters, historic photographs, maps and more. 528pp. 6⅛ x 9¼. 28587-1

ANCIENT EGYPTIAN MATERIALS AND INDUSTRIES, A. Lucas and J. Harris. Fascinating, comprehensive, thoroughly documented text describes this ancient civilization's vast resources and the processes that incorporated them in daily life, including the use of animal products, building materials, cosmetics, perfumes and incense, fibers, glazed ware, glass and its manufacture, materials used in the mummification process, and much more. 544pp. 6⅛ x 9¼. (Available in U.S. only.) 40446-3

RUSSIAN STORIES/RUSSKIE RASSKAZY: A Dual-Language Book, edited by Gleb Struve. Twelve tales by such masters as Chekhov, Tolstoy, Dostoevsky, Pushkin, others. Excellent word-for-word English translations on facing pages, plus teaching and study aids, Russian/English vocabulary, biographical/critical introductions, more. 416pp. 5⅜ x 8½. 26244-8

PHILADELPHIA THEN AND NOW: 60 Sites Photographed in the Past and Present, Kenneth Finkel and Susan Oyama. Rare photographs of City Hall, Logan Square, Independence Hall, Betsy Ross House, other landmarks juxtaposed with contemporary views. Captures changing face of historic city. Introduction. Captions. 128pp. 8¼ x 11. 25790-8

AIA ARCHITECTURAL GUIDE TO NASSAU AND SUFFOLK COUNTIES, LONG ISLAND, The American Institute of Architects, Long Island Chapter, and the Society for the Preservation of Long Island Antiquities. Comprehensive, well-researched and generously illustrated volume brings to life over three centuries of Long Island's great architectural heritage. More than 240 photographs with authoritative, extensively detailed captions. 176pp. 8¼ x 11. 26946-9

NORTH AMERICAN INDIAN LIFE: Customs and Traditions of 23 Tribes, Elsie Clews Parsons (ed.). 27 fictionalized essays by noted anthropologists examine religion, customs, government, additional facets of life among the Winnebago, Crow, Zuni, Eskimo, other tribes. 480pp. 6⅛ x 9¼. 27377-6

FRANK LLOYD WRIGHT'S DANA HOUSE, Donald Hoffmann. Pictorial essay of residential masterpiece with over 160 interior and exterior photos, plans, elevations, sketches and studies. 128pp. 9¼ x 10¾. 29120-0

THE MALE AND FEMALE FIGURE IN MOTION: 60 Classic Photographic Sequences, Eadweard Muybridge. 60 true-action photographs of men and women walking, running, climbing, bending, turning, etc., reproduced from rare 19th-century masterpiece. vi + 121pp. 9 x 12. 24745-7

1001 QUESTIONS ANSWERED ABOUT THE SEASHORE, N. J. Berrill and Jacquelyn Berrill. Queries answered about dolphins, sea snails, sponges, starfish, fishes, shore birds, many others. Covers appearance, breeding, growth, feeding, much more. 305pp. 5¼ x 8¼. 23366-9

ATTRACTING BIRDS TO YOUR YARD, William J. Weber. Easy-to-follow guide offers advice on how to attract the greatest diversity of birds: birdhouses, feeders, water and waterers, much more. 96pp. 5³⁄₁₆ x 8¼. 28927-3

MEDICINAL AND OTHER USES OF NORTH AMERICAN PLANTS: A Historical Survey with Special Reference to the Eastern Indian Tribes, Charlotte Erichsen-Brown. Chronological historical citations document 500 years of usage of plants, trees, shrubs native to eastern Canada, northeastern U.S. Also complete identifying information. 343 illustrations. 544pp. 6½ x 9¼. 25951-X

STORYBOOK MAZES, Dave Phillips. 23 stories and mazes on two-page spreads: Wizard of Oz, Treasure Island, Robin Hood, etc. Solutions. 64pp. 8¼ x 11. 23628-5

AMERICAN NEGRO SONGS: 230 Folk Songs and Spirituals, Religious and Secular, John W. Work. This authoritative study traces the African influences of songs sung and played by black Americans at work, in church, and as entertainment. The author discusses the lyric significance of such songs as "Swing Low, Sweet Chariot," "John Henry," and others and offers the words and music for 230 songs. Bibliography. Index of Song Titles. 272pp. 6½ x 9¼. 40271-1

MOVIE-STAR PORTRAITS OF THE FORTIES, John Kobal (ed.). 163 glamor, studio photos of 106 stars of the 1940s: Rita Hayworth, Ava Gardner, Marlon Brando, Clark Gable, many more. 176pp. 8⅜ x 11¼. 23546-7

BENCHLEY LOST AND FOUND, Robert Benchley. Finest humor from early 30s, about pet peeves, child psychologists, post office and others. Mostly unavailable elsewhere. 73 illustrations by Peter Arno and others. 183pp. 5⅜ x 8½. 22410-4

YEKL and THE IMPORTED BRIDEGROOM AND OTHER STORIES OF YIDDISH NEW YORK, Abraham Cahan. Film Hester Street based on *Yekl* (1896). Novel, other stories among first about Jewish immigrants on N.Y.'s East Side. 240pp. 5⅜ x 8½. 22427-9

SELECTED POEMS, Walt Whitman. Generous sampling from *Leaves of Grass*. Twenty-four poems include "I Hear America Singing," "Song of the Open Road," "I Sing the Body Electric," "When Lilacs Last in the Dooryard Bloom'd," "O Captain! My Captain!"–all reprinted from an authoritative edition. Lists of titles and first lines. 128pp. 5³⁄₁₆ x 8¼. 26878-0

THE BEST TALES OF HOFFMANN, E. T. A. Hoffmann. 10 of Hoffmann's most important stories: "Nutcracker and the King of Mice," "The Golden Flowerpot," etc. 458pp. 5⅜ x 8½. 21793-0

FROM FETISH TO GOD IN ANCIENT EGYPT, E. A. Wallis Budge. Rich detailed survey of Egyptian conception of "God" and gods, magic, cult of animals, Osiris, more. Also, superb English translations of hymns and legends. 240 illustrations. 545pp. 5⅜ x 8½. 25803-3

FRENCH STORIES/CONTES FRANÇAIS: A Dual-Language Book, Wallace Fowlie. Ten stories by French masters, Voltaire to Camus: "Micromegas" by Voltaire; "The Atheist's Mass" by Balzac; "Minuet" by de Maupassant; "The Guest" by Camus, six more. Excellent English translations on facing pages. Also French-English vocabulary list, exercises, more. 352pp. 5⅜ x 8½. 26443-2

CHICAGO AT THE TURN OF THE CENTURY IN PHOTOGRAPHS: 122 Historic Views from the Collections of the Chicago Historical Society, Larry A. Viskochil. Rare large-format prints offer detailed views of City Hall, State Street, the Loop, Hull House, Union Station, many other landmarks, circa 1904-1913. Introduction. Captions. Maps. 144pp. 9⅜ x 12¼. 24656-6

OLD BROOKLYN IN EARLY PHOTOGRAPHS, 1865-1929, William Lee Younger. Luna Park, Gravesend race track, construction of Grand Army Plaza, moving of Hotel Brighton, etc. 157 previously unpublished photographs. 165pp. 8⅞ x 11¾. 23587-4

THE MYTHS OF THE NORTH AMERICAN INDIANS, Lewis Spence. Rich anthology of the myths and legends of the Algonquins, Iroquois, Pawnees and Sioux, prefaced by an extensive historical and ethnological commentary. 36 illustrations. 480pp. 5⅜ x 8½. 25967-6

AN ENCYCLOPEDIA OF BATTLES: Accounts of Over 1,560 Battles from 1479 B.C. to the Present, David Eggenberger. Essential details of every major battle in recorded history from the first battle of Megiddo in 1479 B.C. to Grenada in 1984. List of Battle Maps. New Appendix covering the years 1967-1984. Index. 99 illustrations. 544pp. 6½ x 9¼. 24913-1

SAILING ALONE AROUND THE WORLD, Captain Joshua Slocum. First man to sail around the world, alone, in small boat. One of great feats of seamanship told in delightful manner. 67 illustrations. 294pp. 5⅜ x 8½. 20326-3

ANARCHISM AND OTHER ESSAYS, Emma Goldman. Powerful, penetrating, prophetic essays on direct action, role of minorities, prison reform, puritan hypocrisy, violence, etc. 271pp. 5⅜ x 8½. 22484-8

MYTHS OF THE HINDUS AND BUDDHISTS, Ananda K. Coomaraswamy and Sister Nivedita. Great stories of the epics; deeds of Krishna, Shiva, taken from puranas, Vedas, folk tales; etc. 32 illustrations. 400pp. 5⅜ x 8½. 21759-0

THE TRAUMA OF BIRTH, Otto Rank. Rank's controversial thesis that anxiety neurosis is caused by profound psychological trauma which occurs at birth. 256pp. 5⅜ x 8½. 27974-X

A THEOLOGICO-POLITICAL TREATISE, Benedict Spinoza. Also contains unfinished Political Treatise. Great classic on religious liberty, theory of government on common consent. R. Elwes translation. Total of 421pp. 5⅜ x 8½. 20249-6

MY BONDAGE AND MY FREEDOM, Frederick Douglass. Born a slave, Douglass became outspoken force in antislavery movement. The best of Douglass' autobiographies. Graphic description of slave life. 464pp. 5⅜ x 8½. 22457-0

FOLLOWING THE EQUATOR: A Journey Around the World, Mark Twain. Fascinating humorous account of 1897 voyage to Hawaii, Australia, India, New Zealand, etc. Ironic, bemused reports on peoples, customs, climate, flora and fauna, politics, much more. 197 illustrations. 720pp. 5⅜ x 8½. 26113-1

THE PEOPLE CALLED SHAKERS, Edward D. Andrews. Definitive study of Shakers: origins, beliefs, practices, dances, social organization, furniture and crafts, etc. 33 illustrations. 351pp. 5⅜ x 8½. 21081-2

THE MYTHS OF GREECE AND ROME, H. A. Guerber. A classic of mythology, generously illustrated, long prized for its simple, graphic, accurate retelling of the principal myths of Greece and Rome, and for its commentary on their origins and significance. With 64 illustrations by Michelangelo, Raphael, Titian, Rubens, Canova, Bernini and others. 480pp. 5⅜ x 8½. 27584-1

PSYCHOLOGY OF MUSIC, Carl E. Seashore. Classic work discusses music as a medium from psychological viewpoint. Clear treatment of physical acoustics, auditory apparatus, sound perception, development of musical skills, nature of musical feeling, host of other topics. 88 figures. 408pp. 5⅜ x 8½. 21851-1

THE PHILOSOPHY OF HISTORY, Georg W. Hegel. Great classic of Western thought develops concept that history is not chance but rational process, the evolution of freedom. 457pp. 5⅜ x 8½. 20112-0

THE BOOK OF TEA, Kakuzo Okakura. Minor classic of the Orient: entertaining, charming explanation, interpretation of traditional Japanese culture in terms of tea ceremony. 94pp. 5⅜ x 8½. 20070-1

LIFE IN ANCIENT EGYPT, Adolf Erman. Fullest, most thorough, detailed older account with much not in more recent books, domestic life, religion, magic, medicine, commerce, much more. Many illustrations reproduce tomb paintings, carvings, hieroglyphs, etc. 597pp. 5⅜ x 8½. 22632-8

SUNDIALS, Their Theory and Construction, Albert Waugh. Far and away the best, most thorough coverage of ideas, mathematics concerned, types, construction, adjusting anywhere. Simple, nontechnical treatment allows even children to build several of these dials. Over 100 illustrations. 230pp. 5⅜ x 8½. 22947-5

THEORETICAL HYDRODYNAMICS, L. M. Milne-Thomson. Classic exposition of the mathematical theory of fluid motion, applicable to both hydrodynamics and aerodynamics. Over 600 exercises. 768pp. 6⅛ x 9¼. 68970-0

SONGS OF EXPERIENCE: Facsimile Reproduction with 26 Plates in Full Color, William Blake. 26 full-color plates from a rare 1826 edition. Includes "The Tyger," "London," "Holy Thursday," and other poems. Printed text of poems. 48pp. 5¼ x 7. 24636-1

OLD-TIME VIGNETTES IN FULL COLOR, Carol Belanger Grafton (ed.). Over 390 charming, often sentimental illustrations, selected from archives of Victorian graphics—pretty women posing, children playing, food, flowers, kittens and puppies, smiling cherubs, birds and butterflies, much more. All copyright-free. 48pp. 9¼ x 12¼. 27269-9

PERSPECTIVE FOR ARTISTS, Rex Vicat Cole. Depth, perspective of sky and sea, shadows, much more, not usually covered. 391 diagrams, 81 reproductions of drawings and paintings. 279pp. 5⅜ x 8½. 22487-2

DRAWING THE LIVING FIGURE, Joseph Sheppard. Innovative approach to artistic anatomy focuses on specifics of surface anatomy, rather than muscles and bones. Over 170 drawings of live models in front, back and side views, and in widely varying poses. Accompanying diagrams. 177 illustrations. Introduction. Index. 144pp. 8⅜ x11¼. 26723-7

GOTHIC AND OLD ENGLISH ALPHABETS: 100 Complete Fonts, Dan X. Solo. Add power, elegance to posters, signs, other graphics with 100 stunning copyright-free alphabets: Blackstone, Dolbey, Germania, 97 more—including many lower-case, numerals, punctuation marks. 104pp. 8⅛ x 11. 24695-7

HOW TO DO BEADWORK, Mary White. Fundamental book on craft from simple projects to five-bead chains and woven works. 106 illustrations. 142pp. 5⅜ x 8. 20697-1

THE BOOK OF WOOD CARVING, Charles Marshall Sayers. Finest book for beginners discusses fundamentals and offers 34 designs. "Absolutely first rate . . . well thought out and well executed."–E. J. Tangerman. 118pp. 7¾ x 10⅝. 23654-4

ILLUSTRATED CATALOG OF CIVIL WAR MILITARY GOODS: Union Army Weapons, Insignia, Uniform Accessories, and Other Equipment, Schuyler, Hartley, and Graham. Rare, profusely illustrated 1846 catalog includes Union Army uniform and dress regulations, arms and ammunition, coats, insignia, flags, swords, rifles, etc. 226 illustrations. 160pp. 9 x 12. 24939-5

WOMEN'S FASHIONS OF THE EARLY 1900s: An Unabridged Republication of "New York Fashions, 1909," National Cloak & Suit Co. Rare catalog of mail-order fashions documents women's and children's clothing styles shortly after the turn of the century. Captions offer full descriptions, prices. Invaluable resource for fashion, costume historians. Approximately 725 illustrations. 128pp. 8⅜ x 11¼. 27276-1

THE 1912 AND 1915 GUSTAV STICKLEY FURNITURE CATALOGS, Gustav Stickley. With over 200 detailed illustrations and descriptions, these two catalogs are essential reading and reference materials and identification guides for Stickley furniture. Captions cite materials, dimensions and prices. 112pp. 6½ x 9¼. 26676-1

EARLY AMERICAN LOCOMOTIVES, John H. White, Jr. Finest locomotive engravings from early 19th century: historical (1804–74), main-line (after 1870), special, foreign, etc. 147 plates. 142pp. 11⅜ x 8¼. 22772-3

THE TALL SHIPS OF TODAY IN PHOTOGRAPHS, Frank O. Braynard. Lavishly illustrated tribute to nearly 100 majestic contemporary sailing vessels: Amerigo Vespucci, Clearwater, Constitution, Eagle, Mayflower, Sea Cloud, Victory, many more. Authoritative captions provide statistics, background on each ship. 190 black-and-white photographs and illustrations. Introduction. 128pp. 8⅞ x 11¾. 27163-3

LITTLE BOOK OF EARLY AMERICAN CRAFTS AND TRADES, Peter Stockham (ed.). 1807 children's book explains crafts and trades: baker, hatter, cooper, potter, and many others. 23 copperplate illustrations. 140pp. 4⅝ x 6.   23336-7

VICTORIAN FASHIONS AND COSTUMES FROM HARPER'S BAZAR, 1867–1898, Stella Blum (ed.). Day costumes, evening wear, sports clothes, shoes, hats, other accessories in over 1,000 detailed engravings. 320pp. 9⅜ x 12¼.   22990-4

GUSTAV STICKLEY, THE CRAFTSMAN, Mary Ann Smith. Superb study surveys broad scope of Stickley's achievement, especially in architecture. Design philosophy, rise and fall of the Craftsman empire, descriptions and floor plans for many Craftsman houses, more. 86 black-and-white halftones. 31 line illustrations. Introduction 208pp. 6½ x 9¼.   27210-9

THE LONG ISLAND RAIL ROAD IN EARLY PHOTOGRAPHS, Ron Ziel. Over 220 rare photos, informative text document origin (1844) and development of rail service on Long Island. Vintage views of early trains, locomotives, stations, passengers, crews, much more. Captions. 8⅞ x 11¾.   26301-0

VOYAGE OF THE LIBERDADE, Joshua Slocum. Great 19th-century mariner's thrilling, first-hand account of the wreck of his ship off South America, the 35-foot boat he built from the wreckage, and its remarkable voyage home. 128pp. 5⅜ x 8½.   40022-0

TEN BOOKS ON ARCHITECTURE, Vitruvius. The most important book ever written on architecture. Early Roman aesthetics, technology, classical orders, site selection, all other aspects. Morgan translation. 331pp. 5⅜ x 8½.   20645-9

THE HUMAN FIGURE IN MOTION, Eadweard Muybridge. More than 4,500 stopped-action photos, in action series, showing undraped men, women, children jumping, lying down, throwing, sitting, wrestling, carrying, etc. 390pp. 7⅞ x 10⅝.   20204-6 Clothbd.

TREES OF THE EASTERN AND CENTRAL UNITED STATES AND CANADA, William M. Harlow. Best one-volume guide to 140 trees. Full descriptions, woodlore, range, etc. Over 600 illustrations. Handy size. 288pp. 4½ x 6⅜.   20395-6

SONGS OF WESTERN BIRDS, Dr. Donald J. Borror. Complete song and call repertoire of 60 western species, including flycatchers, juncoes, cactus wrens, many more–includes fully illustrated booklet.   Cassette and manual 99913-0

GROWING AND USING HERBS AND SPICES, Milo Miloradovich. Versatile handbook provides all the information needed for cultivation and use of all the herbs and spices available in North America. 4 illustrations. Index. Glossary. 236pp. 5⅜ x 8½.   25058-X

BIG BOOK OF MAZES AND LABYRINTHS, Walter Shepherd. 50 mazes and labyrinths in all–classical, solid, ripple, and more–in one great volume. Perfect inexpensive puzzler for clever youngsters. Full solutions. 112pp. 8⅛ x 11.   22951-3

PIANO TUNING, J. Cree Fischer. Clearest, best book for beginner, amateur. Simple repairs, raising dropped notes, tuning by easy method of flattened fifths. No previous skills needed. 4 illustrations. 201pp. 5⅜ x 8½. 23267-0

HINTS TO SINGERS, Lillian Nordica. Selecting the right teacher, developing confidence, overcoming stage fright, and many other important skills receive thoughtful discussion in this indispensible guide, written by a world-famous diva of four decades' experience. 96pp. 5⅜ x 8½. 40094-8

THE COMPLETE NONSENSE OF EDWARD LEAR, Edward Lear. All nonsense limericks, zany alphabets, Owl and Pussycat, songs, nonsense botany, etc., illustrated by Lear. Total of 320pp. 5⅜ x 8½. (Available in U.S. only.) 20167-8

VICTORIAN PARLOUR POETRY: An Annotated Anthology, Michael R. Turner. 117 gems by Longfellow, Tennyson, Browning, many lesser-known poets. "The Village Blacksmith," "Curfew Must Not Ring Tonight," "Only a Baby Small," dozens more, often difficult to find elsewhere. Index of poets, titles, first lines. xxiii + 325pp. 5⅜ x 8¼. 27044-0

DUBLINERS, James Joyce. Fifteen stories offer vivid, tightly focused observations of the lives of Dublin's poorer classes. At least one, "The Dead," is considered a masterpiece. Reprinted complete and unabridged from standard edition. 160pp. 5³⁄₁₆ x 8¼. 26870-5

GREAT WEIRD TALES: 14 Stories by Lovecraft, Blackwood, Machen and Others, S. T. Joshi (ed.). 14 spellbinding tales, including "The Sin Eater," by Fiona McLeod, "The Eye Above the Mantel," by Frank Belknap Long, as well as renowned works by R. H. Barlow, Lord Dunsany, Arthur Machen, W. C. Morrow and eight other masters of the genre. 256pp. 5⅜ x 8½. (Available in U.S. only.) 40436-6

THE BOOK OF THE SACRED MAGIC OF ABRAMELIN THE MAGE, translated by S. MacGregor Mathers. Medieval manuscript of ceremonial magic. Basic document in Aleister Crowley, Golden Dawn groups. 268pp. 5⅜ x 8½. 23211-5

NEW RUSSIAN-ENGLISH AND ENGLISH-RUSSIAN DICTIONARY, M. A. O'Brien. This is a remarkably handy Russian dictionary, containing a surprising amount of information, including over 70,000 entries. 366pp. 4½ x 6⅛. 20208-9

HISTORIC HOMES OF THE AMERICAN PRESIDENTS, Second, Revised Edition, Irvin Haas. A traveler's guide to American Presidential homes, most open to the public, depicting and describing homes occupied by every American President from George Washington to George Bush. With visiting hours, admission charges, travel routes. 175 photographs. Index. 160pp. 8¼ x 11. 26751-2

NEW YORK IN THE FORTIES, Andreas Feininger. 162 brilliant photographs by the well-known photographer, formerly with *Life* magazine. Commuters, shoppers, Times Square at night, much else from city at its peak. Captions by John von Hartz. 181pp. 9¼ x 10¾. 23585-8

INDIAN SIGN LANGUAGE, William Tomkins. Over 525 signs developed by Sioux and other tribes. Written instructions and diagrams. Also 290 pictographs. 111pp. 6⅛ x 9¼. 22029-X

ANATOMY: A Complete Guide for Artists, Joseph Sheppard. A master of figure drawing shows artists how to render human anatomy convincingly. Over 460 illustrations. 224pp. 8⅜ x 11¼. 27279-6

MEDIEVAL CALLIGRAPHY: Its History and Technique, Marc Drogin. Spirited history, comprehensive instruction manual covers 13 styles (ca. 4th century through 15th). Excellent photographs; directions for duplicating medieval techniques with modern tools. 224pp. 8⅜ x 11¼. 26142-5

DRIED FLOWERS: How to Prepare Them, Sarah Whitlock and Martha Rankin. Complete instructions on how to use silica gel, meal and borax, perlite aggregate, sand and borax, glycerine and water to create attractive permanent flower arrangements. 12 illustrations. 32pp. 5⅜ x 8½. 21802-3

EASY-TO-MAKE BIRD FEEDERS FOR WOODWORKERS, Scott D. Campbell. Detailed, simple-to-use guide for designing, constructing, caring for and using feeders. Text, illustrations for 12 classic and contemporary designs. 96pp. 5⅜ x 8½. 25847-5

SCOTTISH WONDER TALES FROM MYTH AND LEGEND, Donald A. Mackenzie. 16 lively tales tell of giants rumbling down mountainsides, of a magic wand that turns stone pillars into warriors, of gods and goddesses, evil hags, powerful forces and more. 240pp. 5⅜ x 8½. 29677-6

THE HISTORY OF UNDERCLOTHES, C. Willett Cunnington and Phyllis Cunnington. Fascinating, well-documented survey covering six centuries of English undergarments, enhanced with over 100 illustrations: 12th-century laced-up bodice, footed long drawers (1795), 19th-century bustles, 19th-century corsets for men, Victorian "bust improvers," much more. 272pp. 5⅝ x 8¼. 27124-2

ARTS AND CRAFTS FURNITURE: The Complete Brooks Catalog of 1912, Brooks Manufacturing Co. Photos and detailed descriptions of more than 150 now very collectible furniture designs from the Arts and Crafts movement depict davenports, settees, buffets, desks, tables, chairs, bedsteads, dressers and more, all built of solid, quarter-sawed oak. Invaluable for students and enthusiasts of antiques, Americana and the decorative arts. 80pp. 6½ x 9¼. 27471-3

WILBUR AND ORVILLE: A Biography of the Wright Brothers, Fred Howard. Definitive, crisply written study tells the full story of the brothers' lives and work. A vividly written biography, unparalleled in scope and color, that also captures the spirit of an extraordinary era. 560pp. 6⅛ x 9¼. 40297-5

THE ARTS OF THE SAILOR: Knotting, Splicing and Ropework, Hervey Garrett Smith. Indispensable shipboard reference covers tools, basic knots and useful hitches; handsewing and canvas work, more. Over 100 illustrations. Delightful reading for sea lovers. 256pp. 5⅜ x 8½. 26440-8

FRANK LLOYD WRIGHT'S FALLINGWATER: The House and Its History, Second, Revised Edition, Donald Hoffmann. A total revision–both in text and illustrations–of the standard document on Fallingwater, the boldest, most personal architectural statement of Wright's mature years, updated with valuable new material from the recently opened Frank Lloyd Wright Archives. "Fascinating"–*The New York Times*. 116 illustrations. 128pp. 9¼ x 10¾. 27430-6

PHOTOGRAPHIC SKETCHBOOK OF THE CIVIL WAR, Alexander Gardner. 100 photos taken on field during the Civil War. Famous shots of Manassas Harper's Ferry, Lincoln, Richmond, slave pens, etc. 244pp. 10⅝ x 8¼. 22731-6

FIVE ACRES AND INDEPENDENCE, Maurice G. Kains. Great back-to-the-land classic explains basics of self-sufficient farming. The one book to get. 95 illustrations. 397pp. 5⅜ x 8½. 20974-1

SONGS OF EASTERN BIRDS, Dr. Donald J. Borror. Songs and calls of 60 species most common to eastern U.S.: warblers, woodpeckers, flycatchers, thrushes, larks, many more in high-quality recording. Cassette and manual 99912-2

A MODERN HERBAL, Margaret Grieve. Much the fullest, most exact, most useful compilation of herbal material. Gigantic alphabetical encyclopedia, from aconite to zedoary, gives botanical information, medical properties, folklore, economic uses, much else. Indispensable to serious reader. 161 illustrations. 888pp. 6½ x 9¼. 2-vol. set. (Available in U.S. only.) Vol. I: 22798-7
Vol. II: 22799-5

HIDDEN TREASURE MAZE BOOK, Dave Phillips. Solve 34 challenging mazes accompanied by heroic tales of adventure. Evil dragons, people-eating plants, blood-thirsty giants, many more dangerous adversaries lurk at every twist and turn. 34 mazes, stories, solutions. 48pp. 8¼ x 11. 24566-7

LETTERS OF W. A. MOZART, Wolfgang A. Mozart. Remarkable letters show bawdy wit, humor, imagination, musical insights, contemporary musical world; includes some letters from Leopold Mozart. 276pp. 5⅜ x 8½. 22859-2

BASIC PRINCIPLES OF CLASSICAL BALLET, Agrippina Vaganova. Great Russian theoretician, teacher explains methods for teaching classical ballet. 118 illustrations. 175pp. 5⅜ x 8½. 22036-2

THE JUMPING FROG, Mark Twain. Revenge edition. The original story of The Celebrated Jumping Frog of Calaveras County, a hapless French translation, and Twain's hilarious "retranslation" from the French. 12 illustrations. 66pp. 5⅜ x 8½. 22686-7

BEST REMEMBERED POEMS, Martin Gardner (ed.). The 126 poems in this superb collection of 19th- and 20th-century British and American verse range from Shelley's "To a Skylark" to the impassioned "Renascence" of Edna St. Vincent Millay and to Edward Lear's whimsical "The Owl and the Pussycat." 224pp. 5⅜ x 8½. 27165-X

COMPLETE SONNETS, William Shakespeare. Over 150 exquisite poems deal with love, friendship, the tyranny of time, beauty's evanescence, death and other themes in language of remarkable power, precision and beauty. Glossary of archaic terms. 80pp. 5³⁄₁₆ x 8¼. 26686-9

THE BATTLES THAT CHANGED HISTORY, Fletcher Pratt. Eminent historian profiles 16 crucial conflicts, ancient to modern, that changed the course of civilization. 352pp. 5⅜ x 8½. 41129-X

THE WIT AND HUMOR OF OSCAR WILDE, Alvin Redman (ed.). More than 1,000 ripostes, paradoxes, wisecracks: Work is the curse of the drinking classes; I can resist everything except temptation; etc. 258pp. 5⅜ x 8½. 20602-5

SHAKESPEARE LEXICON AND QUOTATION DICTIONARY, Alexander Schmidt. Full definitions, locations, shades of meaning in every word in plays and poems. More than 50,000 exact quotations. 1,485pp. 6½ x 9¼. 2-vol. set.
Vol. 1: 22726-X
Vol. 2: 22727-8

SELECTED POEMS, Emily Dickinson. Over 100 best-known, best-loved poems by one of America's foremost poets, reprinted from authoritative early editions. No comparable edition at this price. Index of first lines. 64pp. 5³⁄₁₆ x 8¼. 26466-1

THE INSIDIOUS DR. FU-MANCHU, Sax Rohmer. The first of the popular mystery series introduces a pair of English detectives to their archnemesis, the diabolical Dr. Fu-Manchu. Flavorful atmosphere, fast-paced action, and colorful characters enliven this classic of the genre. 208pp. 5³⁄₁₆ x 8¼. 29898-1

THE MALLEUS MALEFICARUM OF KRAMER AND SPRENGER, translated by Montague Summers. Full text of most important witchhunter's "bible," used by both Catholics and Protestants. 278pp. 6⅝ x 10. 22802-9

SPANISH STORIES/CUENTOS ESPAÑOLES: A Dual-Language Book, Angel Flores (ed.). Unique format offers 13 great stories in Spanish by Cervantes, Borges, others. Faithful English translations on facing pages. 352pp. 5⅜ x 8½. 25399-6

GARDEN CITY, LONG ISLAND, IN EARLY PHOTOGRAPHS, 1869–1919, Mildred H. Smith. Handsome treasury of 118 vintage pictures, accompanied by carefully researched captions, document the Garden City Hotel fire (1899), the Vanderbilt Cup Race (1908), the first airmail flight departing from the Nassau Boulevard Aerodrome (1911), and much more. 96pp. 8⅞ x 11¾. 40669-5

OLD QUEENS, N.Y., IN EARLY PHOTOGRAPHS, Vincent F. Seyfried and William Asadorian. Over 160 rare photographs of Maspeth, Jamaica, Jackson Heights, and other areas. Vintage views of DeWitt Clinton mansion, 1939 World's Fair and more. Captions. 192pp. 8⅞ x 11. 26358-4

CAPTURED BY THE INDIANS: 15 Firsthand Accounts, 1750-1870, Frederick Drimmer. Astounding true historical accounts of grisly torture, bloody conflicts, relentless pursuits, miraculous escapes and more, by people who lived to tell the tale. 384pp. 5⅜ x 8½. 24901-8

THE WORLD'S GREAT SPEECHES (Fourth Enlarged Edition), Lewis Copeland, Lawrence W. Lamm, and Stephen J. McKenna. Nearly 300 speeches provide public speakers with a wealth of updated quotes and inspiration–from Pericles' funeral oration and William Jennings Bryan's "Cross of Gold Speech" to Malcolm X's powerful words on the Black Revolution and Earl of Spenser's tribute to his sister, Diana, Princess of Wales. 944pp. 5⅜ x 8⅜. 40903-1

THE BOOK OF THE SWORD, Sir Richard F. Burton. Great Victorian scholar/adventurer's eloquent, erudite history of the "queen of weapons"–from prehistory to early Roman Empire. Evolution and development of early swords, variations (sabre, broadsword, cutlass, scimitar, etc.), much more. 336pp. 6⅛ x 9¼.
25434-8

AUTOBIOGRAPHY: The Story of My Experiments with Truth, Mohandas K. Gandhi. Boyhood, legal studies, purification, the growth of the Satyagraha (nonviolent protest) movement. Critical, inspiring work of the man responsible for the freedom of India. 480pp. 5⅜ x 8½. (Available in U.S. only.) 24593-4

CELTIC MYTHS AND LEGENDS, T. W. Rolleston. Masterful retelling of Irish and Welsh stories and tales. Cuchulain, King Arthur, Deirdre, the Grail, many more. First paperback edition. 58 full-page illustrations. 512pp. 5⅜ x 8½. 26507-2

THE PRINCIPLES OF PSYCHOLOGY, William James. Famous long course complete, unabridged. Stream of thought, time perception, memory, experimental methods; great work decades ahead of its time. 94 figures. 1,391pp. 5⅜ x 8½. 2-vol. set.
Vol. I: 20381-6   Vol. II: 20382-4

THE WORLD AS WILL AND REPRESENTATION, Arthur Schopenhauer. Definitive English translation of Schopenhauer's life work, correcting more than 1,000 errors, omissions in earlier translations. Translated by E. F. J. Payne. Total of 1,269pp. 5⅜ x 8½. 2-vol. set.
Vol. 1: 21761-2   Vol. 2: 21762-0

MAGIC AND MYSTERY IN TIBET, Madame Alexandra David-Neel. Experiences among lamas, magicians, sages, sorcerers, Bonpa wizards. A true psychic discovery. 32 illustrations. 321pp. 5⅜ x 8½. (Available in U.S. only.) 22682-4

THE EGYPTIAN BOOK OF THE DEAD, E. A. Wallis Budge. Complete reproduction of Ani's papyrus, finest ever found. Full hieroglyphic text, interlinear transliteration, word-for-word translation, smooth translation. 533pp. 6½ x 9¼. 21866-X

MATHEMATICS FOR THE NONMATHEMATICIAN, Morris Kline. Detailed, college-level treatment of mathematics in cultural and historical context, with numerous exercises. Recommended Reading Lists. Tables. Numerous figures. 641pp. 5⅜ x 8½. 24823-2

PROBABILISTIC METHODS IN THE THEORY OF STRUCTURES, Isaac Elishakoff. Well-written introduction covers the elements of the theory of probability from two or more random variables, the reliability of such multivariable structures, the theory of random function, Monte Carlo methods of treating problems incapable of exact solution, and more. Examples. 502pp. 5⅜ x 8½. 40691-1

THE RIME OF THE ANCIENT MARINER, Gustave Doré, S. T. Coleridge. Doré's finest work; 34 plates capture moods, subtleties of poem. Flawless full-size reproductions printed on facing pages with authoritative text of poem. "Beautiful. Simply beautiful."–*Publisher's Weekly.* 77pp. 9¼ x 12. 22305-1

NORTH AMERICAN INDIAN DESIGNS FOR ARTISTS AND CRAFTSPEOPLE, Eva Wilson. Over 360 authentic copyright-free designs adapted from Navajo blankets, Hopi pottery, Sioux buffalo hides, more. Geometrics, symbolic figures, plant and animal motifs, etc. 128pp. 8⅜ x 11. (Not for sale in the United Kingdom.) 25341-4

SCULPTURE: Principles and Practice, Louis Slobodkin. Step-by-step approach to clay, plaster, metals, stone; classical and modern. 253 drawings, photos. 255pp. 8⅛ x 11. 22960-2

THE INFLUENCE OF SEA POWER UPON HISTORY, 1660–1783, A. T. Mahan. Influential classic of naval history and tactics still used as text in war colleges. First paperback edition. 4 maps. 24 battle plans. 640pp. 5⅜ x 8½. 25509-3

THE STORY OF THE TITANIC AS TOLD BY ITS SURVIVORS, Jack Winocour (ed.). What it was really like. Panic, despair, shocking inefficiency, and a little heroism. More thrilling than any fictional account. 26 illustrations. 320pp. 5⅜ x 8½. 20610-6

FAIRY AND FOLK TALES OF THE IRISH PEASANTRY, William Butler Yeats (ed.). Treasury of 64 tales from the twilight world of Celtic myth and legend: "The Soul Cages," "The Kildare Pooka," "King O'Toole and his Goose," many more. Introduction and Notes by W. B. Yeats. 352pp. 5⅜ x 8½. 26941-8

BUDDHIST MAHAYANA TEXTS, E. B. Cowell and others (eds.). Superb, accurate translations of basic documents in Mahayana Buddhism, highly important in history of religions. The Buddha-karita of Asvaghosha, Larger Sukhavativyuha, more. 448pp. 5⅜ x 8½. 25552-2

ONE TWO THREE . . . INFINITY: Facts and Speculations of Science, George Gamow. Great physicist's fascinating, readable overview of contemporary science: number theory, relativity, fourth dimension, entropy, genes, atomic structure, much more. 128 illustrations. Index. 352pp. 5⅜ x 8½. 25664-2

EXPERIMENTATION AND MEASUREMENT, W. J. Youden. Introductory manual explains laws of measurement in simple terms and offers tips for achieving accuracy and minimizing errors. Mathematics of measurement, use of instruments, experimenting with machines. 1994 edition. Foreword. Preface. Introduction. Epilogue. Selected Readings. Glossary. Index. Tables and figures. 128pp. 5⅜ x 8½. 40451-X

DALÍ ON MODERN ART: The Cuckolds of Antiquated Modern Art, Salvador Dalí. Influential painter skewers modern art and its practitioners. Outrageous evaluations of Picasso, Cézanne, Turner, more. 15 renderings of paintings discussed. 44 calligraphic decorations by Dalí. 96pp. 5⅜ x 8½. (Available in U.S. only.) 29220-7

ANTIQUE PLAYING CARDS: A Pictorial History, Henry René D'Allemagne. Over 900 elaborate, decorative images from rare playing cards (14th–20th centuries): Bacchus, death, dancing dogs, hunting scenes, royal coats of arms, players cheating, much more. 96pp. 9¼ x 12¼. 29265-7

MAKING FURNITURE MASTERPIECES: 30 Projects with Measured Drawings, Franklin H. Gottshall. Step-by-step instructions, illustrations for constructing handsome, useful pieces, among them a Sheraton desk, Chippendale chair, Spanish desk, Queen Anne table and a William and Mary dressing mirror. 224pp. 8⅛ x 11¼. 29338-6

THE FOSSIL BOOK: A Record of Prehistoric Life, Patricia V. Rich et al. Profusely illustrated definitive guide covers everything from single-celled organisms and dinosaurs to birds and mammals and the interplay between climate and man. Over 1,500 illustrations. 760pp. 7½ x 10⅛. 29371-8

Paperbound unless otherwise indicated. Available at your book dealer, online at **www.doverpublications.com**, or by writing to Dept. GI, Dover Publications, Inc., 31 East 2nd Street, Mineola, NY 11501. For current price information or for free catalogues (please indicate field of interest), write to Dover Publications or log on to **www.doverpublications.com** and see every Dover book in print. Dover publishes more than 500 books each year on science, elementary and advanced mathematics, biology, music, art, literary history, social sciences, and other areas.